D1234868

Collecting
& Repairing
WATCHES

MAX CUTMORE

David & Charles

Other books by Max Cutmore
The Watch Collector's Handbook
The Pocket Watch Handbook
Watches 1850 –1980
Pin Lever Watches

A DAVID & CHARLES BOOK

Originally published as *The Watch Collector's Handbook*
Copyright © Max Cutmore 1976

This edition first published in the UK in 1999

Copyright © Max Cutmore 1999

A catalogue record for this book is available from the British Library.

ISBN 0 7153 0819 X

681.113

Designed by Martin Harris Creative Media

Printed in Great Britain by Tanner & Butler Ltd.

for David & Charles
Brunel House
Newton Abbot   **3 6626 10168 574 1**
Devon

Contents

Title page: An English table-roller lever watch by Wymark, London, no 3898, in a double-bottomed silver case hallmarked 1827. It has a fusee with maintaining power and an oversprung balance. The movement pillars are cylindrical and the movement has a dust ring. The dial is silver with raised gold numerals.

INTRODUCTION

After many more years of experience of watch collecting and research, it seemed proper to revise a book written nearly 25 years ago. Much has changed in that period, including the price and availability of some types of watch. This is probably due to the increased interest in antique items of all kinds fostered by television programmes.

Watches to me are magnetic, but my main interest lies in the watch movement, because the case is simply a 'box' to keep the 'timekeeper' clean and safe from damage. This has led to many of my purchases being uncased movements, and in due course I disposed of all my watches except my first (described in Chapter 4) and my last purchase, the John Bull shown on the jacket of this book. This dramatic change was brought about by the offer of a tea chest full of movements, many of which were incomplete, largely from the period 1880–1960. Since the few movements I already owned were virtually all made before 1880, I have had an interesting time assessing this new period of study. The movements have little financial value but are important in tracing the development of watchmaking from a craft to a factory industry. Fortunately, they include examples of Swiss, American and English factory-made movements. I also find that every time I look at the remaining unorganised items I seem to find some new variety to excite me – not for value, but for novelty.

Prior to disposing of my original collection, I photographed all my watches. The camera has been a valuable asset because, with the owners' permission, I have also been able to photograph other watches or movements which were not for sale. Photography has also helped with my interest in manufacture, and I have been able to use the camera to enhance my studies of the development of the tools and machinery used in the change from 'handmade' to 'machine-made' movements.

What does this book offer the reader? In the first chapter there is an outline of the history of watches from c.1550 to the present day. The second chapter looks at the technical developments in escapements and timekeeping during the same period; this includes a simple discussion of the electronic watch. Then a completely new third chapter discusses the changing watch manufacturing techniques after 1700 in France, Switzerland, America and England, and provides information on the newer watchmaking countries, including Japan, Russia and China. Multinational companies are also given a mention.

Having dealt with 'factual' events, Chapter 4 discusses watch collecting. Simple repairs are considered in Chapter 5, although this is not a repair manual and other books will be needed if the reader intends to become a competent repairer of all the faults which old watch movements, dials and cases have developed in their working life in various environments.

The book ends with a series of Appendices that include a Technical Glossary: the use of this is a more efficient way of understanding horological terminology than searching through the index or the text. Other useful information presented in the Appendices covers hallmarks, watch sizes, recognised watchmakers, societies to join, collections to visit and places to buy watches.

In the end, however, whatever the book suggests you might do, the most important thing is that you enjoy collecting and your collection as much as I have enjoyed mine.

In this book (particularly in Chapters 4 and 5, which have descriptions of actions) the personal pronoun 'he' has been used throughout. The author is, of course, aware that there are both male and female collectors and repairers, and no offence is intended.

CHAPTER 1

HISTORICAL SURVEY

The history of watchmaking may be divided into periods. These periods are not equal in time, but they represent eras in the development of the various skills involved in making a watch. The first lasts until 1600, the second from 1600 to 1675: this is a period of decoration. Then comes the new era of the balance spring, which may be divided into two periods: 1675–1700 and 1700–75. A period of intense mechanical development follows from 1775 to 1830s and finally the time from 1830 to the present is arbitrarily split at 1900, when the wrist watch starts to replace the pocket watch.

WATCHES BEFORE 1600

Mechanical clocks, in which the driving power is obtained from a suspended weight, date from the first half of the fourteenth century. Even a small clock with such a power source is not suitable as a portable timekeeper, but it was not until the late fifteenth century that the coiled spring was used as an alternative means of power. If a watch is defined as a spring-driven timekeeper that is small enough to be carried on the person, it is only necessary to reduce the size of the portable spring-driven clock for the watch to be born.

Few early watches have survived and it is not possible to state exactly when or where the first was made. The earliest dateable examples are a German watch of 1548, probably by Caspar Werner of Nuremberg, and a French watch of 1551 by Jaques de la Garde. The German watch is in a drum-shaped or tambour case of gilded metal and the French watch in an approximately spherical case. English and Swiss makes do not appear until a quarter of a century later, when it is likely that refugee workers from France provided a stimulus for the local craftsmen. There

is some documentary evidence for watches from Italy and the
Low Countries contemporary with, or even pre-dating, the
German example but none have survived – or, rather, none have
yet been discovered and dated. The earliest watches had movements made of iron or steel in cast or fabricated gilt base-metal
cases, but brass soon replaced the iron for wheels and plates. The
casework remained simple, and by 1600 watchmaking was an
established industry in most of Europe.

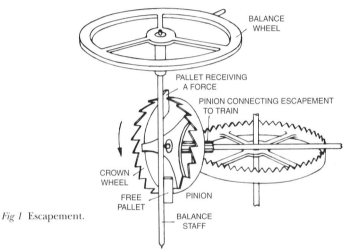

Fig 1 Escapement.

Timekeeping

The timekeeping qualities of early watches are poor. The escapement – the mechanical device by which the spring is allowed to
unwind and so move the hand in small increments – is controlled
by a balance bar or wheel (see Fig 1). The balance rotates first in
one direction and then in the other, under the influence of the
force delivered by the teeth on the crown wheel to the pallets on
the balance staff. The pallets intercept teeth at opposite ends of
the diameter of the crown wheel, so that although the balance
swings to and fro, the rotation of the crown wheel is continuously in one direction. This rotation is passed through the train
(gearing) to allow the hand to rotate at the correct rate and in
the correct direction against the indicating dial. The force delivered by the spring to the train, and hence by the crown wheel to
the balance-staff pallets, depends on the quality and strength of
the spring.

It is not to be expected that early springs would give the same force when fully wound, partly wound or almost unwound. Consequently, the arc of swing of the balance, and therefore the rate at which the train rotates, will vary. If a long spring is used, it is possible to obtain some control of the force delivered by using only the middle portion of the spring. The spring is partly pre-wound or is set up by means of a rachet and pawl, and is stopped from becoming fully wound by positive means; it is hoped that the force delivered by this middle portion is reasonably constant. The amount of set up may be varied with the rachet so that a measure of regulation can be obtained by selecting the portion of spring used. This method of control is inadequate and other devices are required to improve timekeeping.

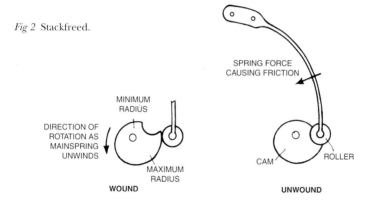

Fig 2 Stackfreed.

An early attempt at further control of the force delivered by the spring was called the stackfreed. In this (see Fig 2), additional friction is introduced to the mechanism by the pressure of an auxiliary leaf spring on a cam. The cam is shaped so that as the driving spring unwinds and the cam rotates, the friction between the auxiliary spring and the cam is reduced to compensate for the reduction in force from the driving spring. The stackfreed is in evidence in early German watches, but French and English makers used a much better device known as a fusee (see Fig 3 overleaf). In this skilful design the spring is contained in a barrel. One end of the spring is attached to the barrel, and the other (inner) end to a fixed arbor about which the barrel rotates. A length of gut (replaced later by chain) is attached, and wound on to, the barrel. The other end of the gut is attached to the fusee.

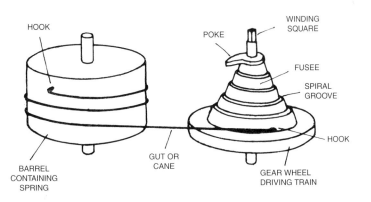

Fig 3 Fusee.

The spring is wound by rotating the fusee, thereby coiling the gut on to the fusee, which has a spiral groove cut on to its cone-shaped profile. When the spring is fully wound, the force which is transmitted to the fusee by the gut acts at a small radius. As the spring unwinds, the gut is recoiled on to the barrel and the reducing spring force acts at the increasing fusee radius. In a correct design the torque (force multiplied by radius) applied to the fusee, and hence to the train, is constant.

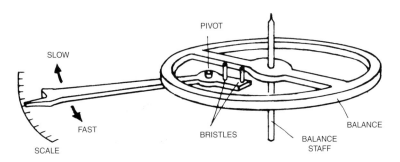

Fig 4 Hog-bristle regulator.

Having made the force delivered by the crown wheel to the balance as constant as possible with set up, stops, stackfreed or fusee, regulation may be achieved by limiting the arc of swing of the balance. The shorter the arc of swing, the quicker the teeth of the crown wheel can escape past the pallets, and the faster the watch will go. In these early watches, the arc of swing was limited by stops which needed some flexibility and were made of hog bristle. The bristles were arranged on a pivoted arm so that their position could be varied (see Fig 4). When the balance arm banks against the bristle the swing is stopped and, under the action of the other pallet on the crown wheel, reverses its direction of rotation.

Even if all these ingenious improvements had functioned perfectly, the friction in the train and in the escapement between the pallets and the crown-wheel teeth would still have varied sufficiently to upset the timekeeping ability of the early watch. This poor timekeeping was apparent in that only a single hand was used to indicate the hour. The metal dial was engraved with numerals at the hours and marks between these to show the half hour. No glass was fitted and the dial often had raised touch pins at the hours to enable the time to be felt in the dark. Striking and alarm work were often included in these early watches: the necessary mechanical arrangements merely needed scaling down from the spring-driven clock. Astronomical data and date mechanisms were also incorporated by the end of the sixteenth century.

Thus, by 1600 the mechanical viability of the watch and the techniques to obtain the best performance were well established in Europe, and there followed a period in which the development of timekeeping was not greatly improved but both the movement and the casework were given decorative attention. Timekeeping improved slightly as the skill of the makers increased. Fusees and springs were better approaches to the ideal and wheels were less hampered by friction. Similarly, the replacement of the ratchet and pawl for set up by a worm and wheel gave finer control, and a small dial indicator was fitted to show the amount of set up used.

1600–75

Early watches were worn on a chain around the neck, and the pendant on the top of the case was usually arranged so that the plane of the hanging ring was at right-angles to the dial of the watch. Since it was not possible to take pride in superior timekeeping, the decoration of the case, which was a visible manifestation of the owner's wealth, became important.

Decorated watch cases were of gilt metal or precious metal and were engraved, jewelled, pierced (especially for striking or alarm) and enamelled. The cases of surviving examples suggest that base metal was common, but this may be because the thick cases required to give rein to the casemakers' skills were, if made of gold or silver, liable to be melted down in hard times in subsequent centuries, when the watch itself was broken or considered inferior as a timekeeper.

The shape of the case changed from the simple tambour cylinder with a lid to a circular or sometimes oval body with hinged domed covers at back and front. There were several exceptions to this generalisation: one such was the form watch, so called because the case was in the form of either a cross, a skull, an animal or a bird; another was the rock-crystal transparent case; and a third the 'puritan' watch in a plain, oval silver case, made in England in about 1640.

Enamel work came in several forms. Champlevé enamel cases were first carved out and the hollows then filled with coloured enamel. Another technique was to paint scenes in enamel on one or more surfaces of the watch covers; and a final variation was high-relief enamel in which the smooth enamel base had raised decorations of leaves or flowers superimposed. Translucent enamel was sometimes used to enhance an engraved case – the underlying design producing a different texture as light was reflected. Sometimes jewels were set in or around enamel scenes to highlight the effect, or the case became just a receptacle for a spectacular display of jewels. From this brief description it is obvious that a watch case would pass through the hands of several craftsmen before the work was complete and the movement finally fitted.

This delicate enamelling and jewelling led to the provision of a protective outer case, often made of leather. To a certain extent the cover defeated the object of the decoration, but it did not have to be worn since the pendant was attached to the inner watch case. If the cover was also worn, it was usually decorated with piqué work in gold or silver pins. A watch with two separate cases, both of which are designed to be worn, is called a pair-case watch. As the pair-case watch developed, the tendency was for the inner case to remain plain and all the decorative effort to be concentrated on the outer cover. The two-case trend coincided to some extent with the change from neck watch to pocket watch, following the introduction of the waistcoat in gentlemen's dress. The hidden pocket watch needed less ostentatious decoration, but it did require the pendant to be rearranged so that the hanging ring was in the same plane as the dial.

Glasses were fitted to cases from about 1620. At first they were held by tabs around the circumference of the hole in the dial cover, and a bezel split at the hinge was introduced a little later. Early glasses were of rock crystal; perhaps the advantage of seeing the time without opening the watch became apparent after this material had been used for some cases. Dials on enamelled case watches were themselves enamelled, usually with decoration or a scene. The hand was made of gold. On metal-cased watches, the dial was usually engraved in the same metal or champlevé with an engraved or matted centre. The hand was of blued steel.

Since the watch owner had to open his watch for winding and regulation, there was also a need to make the movement as attractive as the case and dial. All the brasswork – plates, wheels, pillars and cock – could be gilded. This contrasts well with the blued steel of the set-up worm, springs and screws. The main decorative features were the balance cock and pillars. The balance cock forms the upper bearing for the balance staff and was originally a simple curly C or S design pinned to a block on the plate. As the balance bar disappeared in favour of the balance wheel, the cock became larger to cover the whole wheel protectively. It was circular or oval shaped and was screwed to the plate, being supported by an enlarged foot. Both the cock and the foot were pierced with open scrollwork. The space not used on the plate by cock and foot, set up etc was used to engrave the maker's name and place of business. The pillars separating the plates have considerable variation in shape, becoming progressively more elaborate (see Fig 5). Spiral, round, columnar, baluster, tapering Egyptian

Fig 5 Pillar forms.

SQUARE BALUSTER   ROUND   TULIP   EGYPTIAN

and tulip shape are most common. Additionally, between the plates, the spring winding stop above the fusee was given a decorated pivot.

By 1670 the pocket watch with pair cases had become established and consisted of a decorated outer cover with a plain inner case in gilt metal, silver or gold, or perhaps decorated with enamel, or pierced for striking and alarm. A glass was fitted over the dial. The movement was gilt with worm set up and a gilt pierced cock, decorated pillars and fusee stop pivot, and a replacement for the fusee gut was making an appearance in the form of fine chain with a hook at either end for attachment. The watch was ready for the next important step forward.

1675–1700

The invention of the mainspring as a power source gave birth to the watch in about 1500. The next significant step in its history was the application of the spiral spring to balance control by Huygens in 1675. This single improvement changed the watch from a poor timekeeper with a daily variation measured in scores of minutes to a moderate timekeeper where daily accuracy could be assessed in minutes. There was considerable controversy about this invention both in 1675 and in later times when historians have investigated rival claims. The main argument is whether Hooke or Huygens should be credited with the invention. Both men were involved in experiments with spring control of the balance from about 1660, but Huygens appears to have been the first to apply the spiral form. Some of Hooke's early work was

Fig 6 The principle of Tompion regulation.

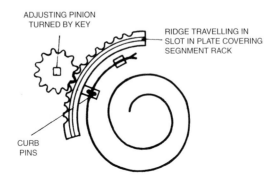

ADJUSTING PINION
TURNED BY KEY

RIDGE TRAVELLING IN
SLOT IN PLATE COVERING
SEGNMENT RACK

CURB
PINS

concerned with straight springs, but it is difficult to reconstruct adequately a time scale for information exchange in those days of poor communication.

There is no doubt that in 1675 Hooke was collaborating with the craftsman Thomas Tompion so as to produce a watch with a spiral balance spring. There is also no doubt that Tompion continued to apply the spiral spring from this date onwards. Tompion gained a reputation as a watchmaker partly as a result of his use of the balance spring, coupled with the use of Hooke's wheel-cutting engine which enabled him to make wheels more accurately, and partly for his superb workmanship. Tompion used a regulator, which enabled the active length of the balance spring to be varied by curb pins operated by a segmental rack (see Fig 6). This allowed the time that the balance took to make a vibration to be adjusted to give accurate timekeeping. Shortening the spring made the watch go faster, while lengthening it made the watch go slower. Tompion was also a good works manager who succeeded in subdividing the work involved in making a watch into skilled and less skilled tasks, so that he could reserve his own efforts for the more important aspects of manufacture.

The action of the balance spring was to use the impulse given to the balance wheel (by the crown wheel through the pallets) to coil up the spring, so bringing the balance to rest. The spring then uncoiled, returning the balance to the centre position. The impulse on the other pallet caused it to uncoil further until the balance again came to rest. Recoiling returned the balance to the initial position. The time taken to complete a full vibration depends on the mass and the mass distribution of the balance, and on the mechanical properties of the spring. In the early days of the application it was hoped that the balance vibration was 'isochronous'– that is, had a time of vibration independent of the arc of vibration. It was found that this was not true, and moreover, it was discovered that the temperature of the watch affected its timekeeping ability, due mainly to the variation of the elasticity of the balance spring with temperature. This fault could only become apparent with the improved timekeeping of the watch. It was also found during this era that the position – pendant up, pendant down etc – of the watch affected its timekeeping. The balance spring opened up a new vista of problems.

Notwithstanding these disadvantages, the improvement in timekeeping achieved by the balance spring was so great that watches were given two hands and dials subdivided into minutes. The dials were still marked with Roman numerals at the hour, but an additional minute ring was provided with slightly smaller Arabic numerals indicating each five minutes around the outer perimeter. Before this method of indication was finalised, other experimental dials and methods were tried (and still are in modern times), but the now conventional arrangement was found to be most suitable.

Regulation by Tompion's method meant that the set up lost its place on the top of the plate and was put between the plates, where it was rarely required except when fitting a new mainspring. The small dial indicator was retained in Tompion's design and used in the same way to determine how the regulator was set. Subsequent to the addition of the balance spring to the watch, a fourth wheel was added to the train. The extra wheel meant that the watch would run for a day between windings rather than the half day achieved by the original three-wheel train.

1700–75

The period from 1700, when the balance spring watch was well established, until 1775 was one in which the English watchmakers were acknowledged to be the best. Disturbing political influences on the Continent aided the skill of the English craftsman to establish this supremacy. Before the turn of the century the mechanism for the repeating watch was perfected independently by Barlow and by Quare, and just after the turn of the century de Facio, a Swiss, together with two emigré Frenchmen (both named Debaufre), took out an English patent for cutting and piercing jewels for bearings.

The repeating watch was made in considerable numbers both in England and on the Continent, but jewelling was confined to England and was rare. Even by 1750 its use was largely restricted to the upper bearing of the balance staff in the cock. Friction in other bearings was reduced by lubrication, and bearing holes in plates were cupped from the middle of the century to provide a small reservoir which restricted the migration of the oil. Dust caps were fitted to movements early in the century, for with the improvement in timekeeping it became noticeable that the intrusion of foreign matter in the movement caused the rate of the watch to deteriorate. The cap had holes in it to allow regulation and winding without the need for complete removal and movement exposure.

Dials in this period at first continued in champlevé but these were slowly replaced by a white, enamelled type with black numerals, so that by 1750 champlevé was rare. The enamel dials continued the practice of Roman hour numerals with Arabic minute indication, but the minute numerals were tending to become smaller as a prelude to their eventual disappearance. Some dials, either of European origin or destined for export, had arched or arcaded chapter rings. Hands were of blued steel or gold in various forms of beetle and poker: the hour hand is known as a beetle hand because of the insect-shaped decoration at its outer end, whereas the minute hand is straight and undecorated and is known as a poker hand.

Cases were relatively simple on the early watches with superior timekeeping, but as this timekeeping ability became the accepted standard and repeating work was not uncommon, case decoration again became fashionable. In this period repoussé outer cases were often in evidence. Repoussé work is achieved by hammering a scene into the metal from inside the case so that a high-relief decoration results. This has a similar look to fine casting. Some enamel work continued but was often restricted to a small scene rather than carried over the whole case. Some outer cases were covered in tortoiseshell, either plain or inlaid with precious metal, while others were covered with translucent horn painted on the underside with fine patterns of ferns or flowers or with scenes. The pendant became more elegant and the hanging ring became a bow, stirrup-shaped and pivoted to the pendant rather than looped through. Glasses were snapped in from early in the century.

Movement decoration continued, in spite of the fact that many were covered with a dust cap and the inner cases were sometimes pierced to allow winding without opening. The cock at first became very large in diameter (balance wheels were larger) with a broad, sector-shaped foot, and much of the rest of the plate was covered with pierced decorative pieces. On the Continent, the

The development of verge-watch style c.1750–1850. All the watches have fusee and chain. *Top left* This watch is by T. Moore, London, no 9291, and is in a silver-gilt inner case hallmarked 1758 with an outer case of underpainted horn. The movement pillars are square baluster. The hands, except for the date indicator, are not original. *Top right* This watch is by Stoakes, London, no 15148, and is in an inner silver case hallmarked 1781 with an outer case of repoussé silver. The movement has square pillars and a bridge cock. The dial is arcaded and the watch is in the 'Dutch forgery' style. *Bottom left* This watch is by W. Hall, London, no 122, and is in silver pair cases hallmarked 1771. The movement has square pillars. The dial is typical of the period, the Arabic numerals outside the Roman numerals becoming less emphasised. The hands are beetle and poker style. *Bottom right* The outer case of the watch above.

cock had no foot but was instead a circular bridge screwed to the plate at either end of a diameter. The early years of the eighteenth century were the peak of movement decoration, and as time passed the foot became smaller and the diameter decreased; both, however, remained pierced. From about 1750 the foot ceased to be pierced and extra decorative pieces became

The makers of these watches are not known. *Left* This watch, in silver pair cases hall-marked 1827, is much larger than the others shown, which is typical of the verge watch of the 1800–30 period. The Arabic numerals on the dial are not as common as Roman numerals. The movement has cylindrical pillars. *Right* This watch is in a silver double bottom case dated 1844. Both watches have verge escapements.

uncommon. Movement decoration never increased again. The small dial indicating the state of the regulation continued throughout the period, and the maker's name and origin were still engraved in the vacant spaces. Pillars became progressively simpler from the elaborate tulip form, and by mid-century small, square baluster pillars were common and some plain cylindrical pillars were in evidence. The decoration of the fusee stop pivot continued, and the end of the spring operating the catch holding the movement in the inner case was often quite elaborate.

Escapements

By 1700 no change in the basic verge escapement of the watch (crown wheel and pallets) had taken place since its inception. No doubt experimental escapements had been tried but none was considered viable. For example, there is a patent of 1695 to Booth (sometimes called Barlow), Houghton and Tompion for a new escapement but no watches have survived. Booth and Houghton were associated with Tompion, as was Graham, and this early patent may have helped Graham to introduce the horizontal or cylinder escapement in 1726. From the timekeeping aspect this escapement was undoubtedly an improvement on the verge, but it was more difficult to make and was fragile. Friction is still present in the design, and the cylinder watch did not displace the verge in England nor even approach it in numbers made. Some eminent makers used it in addition to Graham, who quite naturally concentrated his efforts on the new design.

The cylinder was taken up on the Continent, particularly in France and Switzerland. At the end of this period Lepine used it to produce a much thinner watch, in which the top plate was replaced by a number of bars or bridges. Lepine also dispensed with the fusee and used a going barrel, in which the spring container drove the train directly. The use of stopwork and the improvement in spring quality, combined with the cylinder movement, enabled adequate timekeeping without the fusee.

The smaller, thinner watches were considered more desirable on the Continent, but English makers continued to seek quality by using the fusee in both the verge and the cylinder movement. This could be considered a first step in the decline of the English as the premier watchmakers, but the symptoms were not to be apparent for many years to come.

1775–1830

Geographical position at sea is determined by latitude and longitude. Latitude is quite easily evaluated by observations of the sun or stars, but longitude is more difficult because it is necessary to know the Greenwich time at which the observation is made. The earth takes 24 hours to rotate through 360 degrees, so that a change in longitude of 1 degree is represented by 4 minutes difference in time. For example, the local time of noon is obtained from observations giving the instant of the maximum altitude of the sun. If the Greenwich time of this altitude is one hour after local noon, then the longitude is 15 degrees west of Greenwich. The problem during a voyage is to be sure that the watch or clock indicating Greenwich time is correct or is capable of being corrected. This is not difficult in these modern days of radio communication, but in the eighteenth century the timekeeper could only be set by an observation at a shore station of known longitude. A timekeeper that has a constant rate of loss or gain can be corrected, but one which has a variable rate due to friction, environmental changes etc, cannot be corrected and is useless. The invention of the mainspring made the clock portable and suitable for use at sea; the invention of the balance spring improved timekeeping enormously, but the clock or watch was still not capable of an accuracy suitable for navigation.

In 1714 a huge reward was offered by the Board of Longitude (established by Act of Parliament) to the inventor of a suitable portable timekeeper. The instrument would have to determine longitude to within 30 minutes of arc of a great circle at the end of a voyage from Great Britain to the West Indies to qualify for the maximum award of £20,000. This meant that, after this long voyage, the timekeeper would have to be within two minutes of the correct time, with due allowance for the rate of loss or gain previously stated. The story of the effort to qualify for this award has no place in this review, but with this colossal sum of money available it is not surprising that many attempts were made to construct a timekeeper that would satisfy the requirements.

The successful maker was John Harrison, who spent virtually the whole of his life constructing a series of timekeepers designed to fulfil the requirements of the Board of Longitude. He succeeded with a voyage from Portsmouth to Jamaica taking from 18 November 1761 to 19 January 1762, with a corrected error of one and a quarter minutes of longitude. It took him until 1773, however, with a second demonstration in 1764, to get the full reward.

Although Harrison's marine timekeeper was successful, his contribution to navigation was effectively limited to demonstrating that it was possible to design and construct an accurate device. His mechanisms were too complex. Other eminent horologists, notably Le Roy and Berthoud on the Continent and Arnold and Earnshaw in England, contributed to the production of the practical marine chronometer and, in watch form, the pocket chronometer. Their designs included a spring or pivoted detent escapement, in which the balance is completely detached from the rest of the train (except at the instant it is receiving impulse); a helical balance spring which can be made isochronous more easily than a spiral spring; and maintaining power in the fusee (see Fig 7 overleaf) so that the timekeeper continued to go while

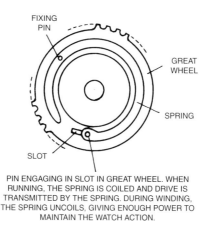

FIXING PIN

GREAT WHEEL

SPRING

SLOT

PIN ENGAGING IN SLOT IN GREAT WHEEL. WHEN
RUNNING, THE SPRING IS COILED AND DRIVE IS
TRANSMITTED BY THE SPRING. DURING WINDING,
THE SPRING UNCOILS, GIVING ENOUGH POWER TO
MAINTAIN THE WATCH ACTION.

Fig 7 Maintaining power.

being wound (Harrison's designs also had maintaining power).
All designs incorporated some form of temperature compensation. The form which survived was the cut, bimetallic balance
and by 1800 the pocket chronometer was a readily available,
accurate watch.

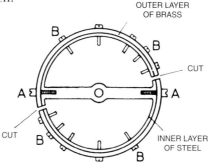

OUTER LAYER
OF BRASS

B   B

CUT

A   A

CUT   B

INNER LAYER
OF STEEL

B

Fig 8 Bimetallic balance wheel.

Some of the improvements were incorporated into the ordinary pocket watch. Maintaining power was used with the fusee
from about 1800, and if the other timekeeping qualities merited
it, a bimetallic balance was fitted (see Fig 8). This design of balance attempts to compensate for the change in elasticity of the
balance spring with temperature by making the balance rim out
of two different metals fused together: thus the outer layer of the

rim is made of brass, which has a higher coefficient of expansion with temperature than the inner layer, which is made of steel. The rim is cut in two places near to the balance arms, so that the free ends will move inward when heated and outward when cooled. This alters the mass distribution of the balance, which will change the time of vibration slightly. Compensation screws are arranged around the rim to give a measure of control to the correction. The timing screws at the ends of the balance arms are not for compensation but are used to adjust the initial mass distribution, so that the fundamental time of vibration is correct. With carefully set timing screws and correct compensation, a regulator is not required and the watch is free sprung. This delicate adjustment is restricted to pocket chronometers and other high-grade watches.

Non-detached escapements

In 1775 the verge watch was still the mainstay of the industry. In the succeeding 50 years a number of escapements were invented or developed from earlier ideas. This period of intense mechanical development was possibly inspired by the success of the chronometer makers, who had demonstrated that accurate time-keeping was possible.

The limitations of the verge watch were a result of the friction in the escapement. The vibration of the balance was always affected by the contact between the crown-wheel teeth and the pallets. Similarly, in the cylinder escapement there was friction between the escape-wheel teeth and the cylinder during the vibration of the balance. To improve the timekeeping qualities of a watch, an escapement was needed in which the balance was detached from the train during the vibration. This was achieved in the detent escapement. Before the development of the detached escapement there were quite a number of other non-detached designs which were tried and which could be considered an improvement on the verge from the timekeeping standpoint, but these were not successful replacements because they were not detached and were harder to make and maintain. The verge is a simple, tough escapement. Three of these escapements with friction were made in sufficient numbers to merit discussion.

The duplex escapement invented in about 1720 by Dutertre,

and modified to its common form around 1750, was used mainly in England from the closing years of the eighteenth century to the middle of the nineteenth century for high-grade watches. The duplex escape wheel has two separate sets of teeth, one for locking the train and one for giving impulse to the balance. It allows escape only once per vibration (as does the detent), and the audible tick is easily recognisable.

An escapement invented by Debaufre in about 1700 has two escape or crown wheels which give impulse to a single inclined plane pallet on the balance staff. It did not achieve great popularity but variations known as the chaffcutter, Ormskirk, club-foot verge and dead-beat verge were developed and made in England in the early nineteenth century. Some of these use two wheels and some two pallets.

Finally, among non-detached escapements there is the rack lever, invented in about 1720 by Abbé de Hautefeuille and improved by Litherland in England in 1791. In the rack lever, the balance staff is fitted with a pinion that is permanently and positively engaged with a rack at the end of a pivoted lever. Impulse is given to an anchor-shaped piece fitted to the lever. The escape wheel has pointed teeth, and the design could be considered as a direct application of the anchor escapement to a watch. This type of escapement can be seen in longcase clocks, but the most common form in the clock has recoil whereas that in the rack lever watch is dead beat. Rack lever watches were made from about 1790 to 1840.

It may be observed that each of these escapements was conceived around 1720, about 40 years after the balance spring had transformed timekeeping, but none of them was sufficient of an improvement on the verge to merit further attention until the period of mechanical interest of 1775–1830. In the intervening time the makers of chronometers had produced a viable, accurate watch, and at some time in the period 1750–60 Mudge had designed the detached lever escapement. Strictly speaking, the Mudge design was not perfectly detached because there was a possibility of friction between the lever and the balance-staff roller. Mudge initially applied his escapement to one or two clocks and is reputed only to have applied it to a single watch in about 1770. He was more interested in the problem of marine timekeeping, and devoted his skill to solving this problem rather than developing the escapement, which was destined to be the solution to cheap and accurate timekeeping.

Several other eminent makers, including Julien le Roy (possibly pre-dating Mudge), Emery, Pendleton, Leroux, Grant, Ellicott, Perigal, Margetts, Robin and Breguet, made a few examples of lever watches in the period up to the turn of the century, but none made it in significant numbers. It seems that the possibility of friction was a stumbling block. Most of these makers changed Mudge's original right-angled layout to a straight-line arrangement (see Fig 9). Peculiarly, the Mudge arrangement was revived and used almost exclusively in the English lever watch, whereas the European makers chose to use the straight-line layout.

Fig 9 Lever escapement layout.

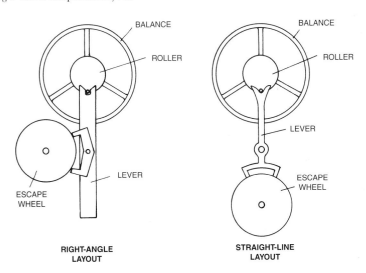

Detached escapements

The missing essential to the lever watch in these early essays was draw. It is not clear who invented draw, in which the escape-wheel teeth are shaped to pull and hold the lever into engagement with the escape wheel thus locking the train, but it was possibly Leroux. The holding of the lever ensured that there was no possibility of slipping into contact with the balance-staff roller, so that the escapement was truly detached and the requirements for good timekeeping (no friction during the vibration) were satisfied.

In 1814 Massey invented a form of lever escapement in which he used the right-angled layout of the rack lever and of Mudge. In his escapement, the rack lever pinion (on the balance staff) is replaced by a single tooth and the rack by a single slot. After the tooth receives impulse from the slot, the balance becomes detached. Some Massey levers do not have draw, so that friction is still possible, while others do. In later forms the single tooth becomes a jewel, supported initially at both top and bottom to form a crank and finally supported only at the top. The escapement in these final forms is also known as the crank lever.

At some time around 1825 the detached table-roller lever watch evolved, when the crank of the Massey lever was replaced by a thin cylindrical roller with an impulse jewel projecting from the base of the cylinder. In this form, with draw, the ultimate replacement for the verge escapement in England had arrived.

A quarter-repeating verge watch by Wm Fredk Strigel, London. The movement is about 1750, but the watch was recased by a worker who wrote inside: 'Strigel, new case, January 18, 1843'. The case is of this period. The bell on which the hammers strike is in the back of the case.

The layout of the English lever watch retained the right-angled arrangement of Mudge, the rack lever and the Massey lever. Continental lever watches employed both the right-angled layout and the straight-line layout. Another variation of the lever appeared in 1815. It was detached and looks similar to the table roller. However, the train is unlocked by two pins on the roller (one for each direction of vibration) and impulse is given to a slot in the roller by a pin on the lever – exactly the reverse of the table roller, in which the slot in the lever gives impulse to the jewel pin on the roller. The construction was more difficult than the table-roller design and the Savage two-pin escapement did not survive.

Other developments

The development of new escapements led to some modification in the layout of the watch. The marked improvement in time-keeping merited the use of the seconds hand and this became conventionally placed, indicating on a small subsidiary dial above the six hour mark. The seconds hand required the train gearing to be arranged so that one wheel rotated at the correct rate and was sited correctly. Jewelling was widely used in the train and some extremely attractive large jewels were fitted on the visible plate. This is sometimes called Liverpool jewelling and indicates the importance achieved by the Lancashire watch industry at this time. Ratchet-and-pawl set up under the dial was used, rather than worm-and-wheel. The plates were arranged in a more con-venient form for assembly and repair. On the dial plate a small bridge bar was used, so that the bottom bearings of the fusee and the fourth wheel (seconds-hand wheel) and the complete third wheel could be fitted after the assembly of escapement and cen-tre wheel. On the other plate the barrel was given a separate bridging plate, so that it could be fitted or the mainspring renewed without the watch having to be completely stripped. Maintaining power was generally used, and on some watches a chronometer-type bimetallic balance was fitted.

Decoration in all watches of this period was becoming mini-mal. The watch was now a timekeeper. Cocks were solid, with some attractive chiselling and engraving and some variety in shape. This may be used to help identify the type of escapement likely to be found between the plates. The cocks were often engraved 'patent' or 'detached'. These varieties were short-lived. Since dust caps were normally fitted, little other decoration was used as only the cock was visible. Pillars were cylindrical and watches tended to be thinner than the 1770 verge, but larger in diameter to accommodate the seconds hand. The regulator was amended to dispense with the segmental rack of Tompion's design and the small indicating dial. Instead there was a simple lever with curb pins embracing the balance spring, which was rotated by hand. The centre of rotation of the lever coincided with the balance axis, so that movement of the lever moved the curb pins around the arc of the balance spring. The end of the lever passed over a scale engraved on the plate to show the amount of regulation. The scale was often marked 'fast' and 'slow' at its ends to make regulation simpler and more accurate.

Many of these improvements and changes were used in the verge watch. In particular, the bridges were added for ease of assembly, and this required the third wheel of the train to be moved to the other side of the contrate wheel. The regulation was made similar to the new, simple design, and the worm set up was replaced by ratchet and pawl under the dial. Set up was usually required only on the rare occasions when the mainspring was renewed, and at this time the dial could be removed to give access. The cock became simpler, with a solid foot and eventually with no piercing. Plain cylindrical pillars were usual, and to keep in style (but not in timekeeping ability), a seconds hand was sometimes fitted. The watch diameter increased substantially but the thickness was not reduced.

Pair cases were not always used in watches with the new escapements, and by 1830 this type of case was rare except on the old-fashioned verge. Where pair cases existed, they were usually of silver or gold with variously shaped pendants and stirrup-shaped bows. Decoration was unusual on either inner or outer case. The alternative single cases used with the new escapement watches were of various designs, but decoration was minimal and usually confined to engine turning on the back and a milled circumference of the centre band. The common form of single case had a double bottom. The watch was wound by opening a hinged back, which revealed a second fixed bottom pierced by a winding hole.

The movement was hinged to the top of the case front as in the inner pair case, but the bezel holding the glass had a separate hinge adjacent to the nine hour mark. Glasses were thinner and flatter to match the larger-diameter flat dial, and the small flat in the centre of the glass disappeared. The pendant was a spherical knob, often pierced by a push piece which released a spring catch holding the back closed. The bow was circular.

Dials were usually white enamel with a single set of numbers indicating the hours. Both Arabic and Roman numerals were used, but Roman continued to be the more common. The dials tended to be less domed because the diameter was greater; dials with seconds hands were flat. Hands became simpler, and the beetle and poker gave way to an hour hand with spade-shaped end and the minute hand was a simple pointer. Seconds hands had a counterpoise with a circular end. The materials used were still gold or blued steel.

The English watch of 1830 was severe but attractive and a good timekeeper. Although a large variety of escapements was available, the English table-roller lever was beginning to become established.

Continental developments

On the Continent, developments were different. It has been explained how the cylinder escapement and the Lepine barred-movement design led to a thinner watch there. In general this trend continued, with wider use of the going barrel rather than the fusee. The fusee was not completely discarded, however, and was retained in many of the traditional verge watches of this period. Verge watches are generally smaller than their English counterparts, often because of the case design. The pair case was abandoned in the quest for minimum size, and often a single bottom case was used, with no dust cap and with winding through the dial. Cocks were small pierced bridges, and the silvered regulator dial continued. The European verge watch presented a very plain and uncluttered movement.

Although the rack lever and duplex were originally European in concept, they did not receive the late-eighteenth-century attention that was given them in England. A variation known as the Chinese duplex was made on the Continent at the end of the period under discussion. In this design, the train was only allowed to move on every other complete balance vibration, and the sweep seconds hand moved forward in uncomfortable-looking steps of one second.

Two further escapements received more attention on the Continent than in England. These were the virgule and the Pouzait. The virgule has an escape wheel with teeth standing up from the plane of the wheel, giving impulse to a comma-shaped piece on the balance staff. The Pouzait is a form of lever escapement which had some following in Switzerland and France. The cylinder watch continued to be made in greater numbers on the Continent than in England but, as in England, there was a move towards the lever.

In 1787 Breguet produced lever watches in France. There is the possibility that the concept of the lever in France by Julien Le

A Continental verge watch of about 1800 in a silver case. The bridge cock, coqueret and regulator figure plate are typical. The movement pillars are five-sided baluster; the crown wheel has adjustable bearings at both ends. The watch is wound through the dial, and there is usually no access through the case back. The movement is fitted with a fusee.

Roy preceded Mudge's first essay, but at the moment there is too little evidence to be certain. There is, however, no doubt about Breguet's work, which is well documented. His design used a straight-line layout with a cut-compensated balance. It is to be expected that anything produced by Breguet would have some innovation and be executed with workmanship of the highest quality – he is probably the most famous European maker noted for his workmanship and style. His best-known work includes ruby cylinder watches, the design of an overcoil for the end of a spiral balance spring giving isochronous vibrations, the shock-absorbing 'parachute' balance staff suspension, a self-winding watch and the tourbillon watch, in which the complete escapement assembly rotates about once every few minutes to minimise positional (pendant up or pendant down) errors in the watch. Breguet's life spanned the period of intense mechanical development and he was no mean contributor.

Apart from the straight-line layout, European lever watches show another important difference from the English design. The English design used an escape wheel with pointed teeth, and the pallets on the working surfaces of the anchor fitted to the lever were steeply angled. This meant that lift-giving impulse to the balance was provided solely by the lever. In the European design the lift was divided between the lever pallets and the escape-wheel teeth, so that these teeth were not pointed but had a broad or club foot.

European cases and dials were in keeping with the slimmer watches being produced. Even in the traditional verge the pair case was not used, and certainly in the new designs the fusee was omitted in the interest of a slim watch in a small, elegant case. Engine turning was used as in England but there was more emphasis on case decoration than in England, where the watch was severe but attractive. European dials were usually white enamel, often with Arabic numerals in greater evidence and with a wider variety of hand forms than would have been found on English watches.

In Switzerland the cylinder was supreme and the makers were steadily moving towards volume production. However, the decisive steps along this path were not taken until the next period, and in 1830 the position on the Continent is summarised by stating that in France the lever was becoming established in straight-line form and in Switzerland the cylinder was poised for its successful one-hundred-year run.

1830–1900

During the first quarter of this period the (English) lever watch became established in England. It was still possible to obtain watches with other escapements, but they became increasingly rare and by 1850 the lever was supreme. Even the verge was faltering after a three-hundred-year history. Changes in the lever watch meant that there was considerable variation in internal and external appearance, and by 1860 the form of the lever itself had changed from the early straight-sided design to a curved one. Many watches were obviously made in the same workshop and then finished by the 'maker' whose name was engraved on the plate. In the traditional full-plate layout, the balance and balance cock are above the plate, giving a thick watch; watches were thinned by using a three-quarter or half-plate movement. In the three-quarter plate, the balance, lever and escape wheel were placed with separate cocks in a space obtained by cutting away a section of the plate. In the half plate the fourth wheel also has a separate cock. The fusee was continued until the last decades of the century – the concessions to thinning had been made.

Winding methods

Changes in the methods of winding occurred continuously and many of the ideas came from the previous era of mechanical activity. In 1814 Massey had produced a push or pump winder with a rack operated by pushing the pendant, turning a ratchet on the fusee or going barrel. In 1820 Prest, who worked for Arnold (the son of the chronometer maker, Arnold), patented a system with a bevel drive through the pendant connected to a going barrel, and in 1827 Berrolas patented a pull-wind system with a recoil spring operating rather like a mower starter. None of these systems was completely satisfactory because a key was still required to adjust the hands. This problem was overcome in some cases by adding a handsetting train operated by a wheel protruding from the side, or back, of the movement. Other variations exist, but the successful winding systems came from the Continent.

The first man to devise winding and handsetting through the pendant was a Swiss maker, Audemars, in 1838. Variations and improvements followed from other Swiss makers (Philippe, Lecoultre and Huguenin) who produced the shifting-sleeve and rocking-bar arrangements. Initially, the change of mode from winding to hand set was by means of a small push piece at the side of the winding button, but this extra piece was not required in the design in which pulling the winding button effected the change of mode.

Fig 10 Keyless winding.

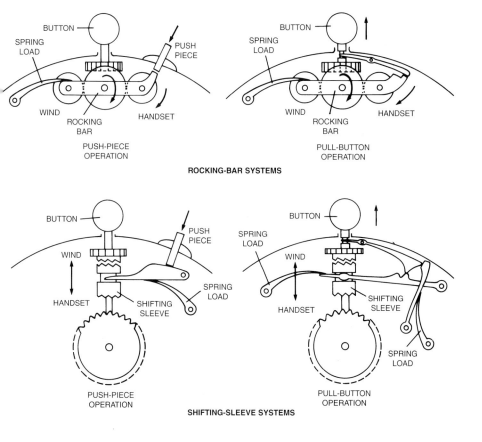

Fig 10 shows four arrangements. There are other variations involving different layouts and the mode to be engaged by button pushing or pulling, and it was many years before the pull-button shifting-sleeve form became the norm. Self-winding watches were conceived by Perrelet (a Swiss) in 1770, and Breguet produced them from 1780, but the 'pedometer' wind did not become a serious pocket-watch device and was discarded until required for wrist watches. Perhaps there was insufficient motion in the average eighteenth-century pocket. (For details on where to find further information on keyless-winding history, see Bibliography.)

The keyless-winding and handsetting system allowed some change in case design. It was not necessary for the front to hinge open for handsetting and so a snap-on bezel was introduced. Similarly, it was only necessary to open the back for regulation or servicing, and as this was normally the job of a watchmaker, the double bottom to the case was abandoned. Instead, the hinged back was snapped firmly shut with a small lifting piece to allow it to be opened. The dust cap on the movement was no longer required and was replaced by a small, hinged cover fitted inside the case back. The movement was no longer hinged to the case but was held in place by screws. When the bezel and screws were removed, the movement could be passed out through the front of the case.

In England, keyless winding was rare in this period, mainly because most of the systems discussed here were only suited for use with the going-barrel watch. In the 1890s there was some tendency to change from the fusee to the going barrel, but the English makers still continued to favour the traditional key-wind watch. The internal style of the watches keeping to the full-plate layout changed little, except that the cocks now reached a minimum of decoration and there were one or two variations in the regulation layout. Liverpool jewelling gave way to less obvious small jewels.

Other developments

The external style changed as the Victorian period progressed. The case and dial became heavier to the eye, losing the austere elegance of 1830. Dials at that time were usually flat and a number of recognisable styles evolved, resulting in the familiar Roman numerals with recessed seconds hand. Hands became slimmer. The quality of the English lever watch of the middle-

An English table-roller lever watch by J. D. Williams, Merthyr, no 27353, in a double-bottomed silver case hallmarked 1873. This is a typical mid-Victorian lever watch of moderate quality, with a cut-compensated balance and a fusee with maintaining power.

and late-Victorian period, taking both timekeeping and workmanship into consideration, was good and such a watch is still capable of giving satisfactory daily performance a century later.

The use of the seconds hand caused stop mechanisms to be added to some watches. In the early designs, a simple internal stop piece operated on the balance rim or the lever. Later in the period there was a group of watches often marked, quite incorrectly, 'chronograph'. In this group the dials were divided in quarters or fifths of seconds, and the rate of vibration was increased to enable these divisions to be realistic; but when the sliding stop piece on the case was moved, the same internal arrangement stopped not only the sweep seconds hand but the whole watch. Timing techniques with these watches were crude. There were more expensive watches with separate trains for the seconds hand and the normal mean-time hands, and these were much better. However, a true chronograph should be capable of being stopped, reset and restarted while the mean-time hands continue. (Similarly, a stop watch should have the stop, reset and restart facility but without mean-time indication.) This capability was designed in 1844 by Nicole, but it was a further 18 years before the familiar three-push system on the winding button was devised. The all-important resetting was achieved by a heart-shaped cam moved by the pressure of a spring. The single-train chronograph drove the seconds hands by friction wheel, engaged and disengaged by the push button.

Left This Continental watch has a cylinder escapement and plated brass case. Although the dial is labelled 'chronograph', it is not, as the stop piece locks the train and stops the watch; there is no reset and start action. *Right* This watch is a chronograph with separate stop, reset and start action for the centre seconds hand. The mean-time hands continue even when the stop is operated. The watch has a Swiss lever escapement and is in a silver case of about 1920. Both watches have a going barrel.

Towards the end of the period all the varieties – repeaters, moonwork, alarm, striking, musical, automata, jaquemarts, multi-dial, day, date, month, stop – were available, and many watches had more than one of these complexities. A large proportion of these watches were Swiss with lever or cylinder escapement, and many had decorated cases. One English refinement was the karrusel, patented in 1892 by Bonniksen. This was a simpler version of the Breguet tourbillon, which compensated for positional error.

Switzerland and America

So far, the most significant happenings of the period, which were to outdate the traditional English watchmaking techniques and see the establishment of the Swiss and American machine-made watch industry, have been bypassed. It is difficult to understand how the Swiss and Americans could see the market potential of cheap machine-made watches when England, where the industrial revolution was born, could not. In particular, the annual export figures achieved in the mid-twentieth century by Switzerland (24 million in 1950, 41 million in 1960 and 63 million in 1967) show the magnitude of the missed opportunity.

From about 1840 on, machine-made watches with some interchangeable parts were produced in Switzerland. The designer of the machines to achieve this milestone was Leschot and his work is described in Chapter 3. The idea of the machine-made watch with interchangeable parts was further developed in America, where there was no traditional industry, and the Americans developed a production system which in turn influenced the Swiss makers.

The way in which these ideas travelled across the world was new. In 1851 the Great Exhibition was held in London, and this was followed by exhibitions in Paris in 1855, London in 1862, Paris in 1867 and Philadelphia in 1876. Further exhibitions followed, but it was the Philadelphia exhibition which enabled the Swiss to realise the magnitude of the American challenge and to take appropriate remedial action. Thus in Europe, the Swiss applied volume-production methods to both the cheap cylinder and the more expensive lever watches, in small fob form for ladies and pocket form for gentlemen. They were sold in ever-increasing numbers throughout the world, and gained new customers as a result of their attendance and winning of medals at the various exhibitions.

The American watch industry is a post-1850 phenomenon. Before this date few watches were made in America – most were imported from Europe, either as parts for assembly or as complete watches. Harland made around two hundred watches in about 1800, and then tariff restrictions in the period 1809–15 encouraged Goddard to make watches; but as soon as the restrictions were removed European imports again dominated the market. Goddard is reputed to have made about five hundred watches. Following this, in 1837–8 the Pitkin brothers produced produced about five hundred machine-made watches with frames that had a degree of interchangeability. Unfortunately for the brothers, the jewellery trade did not accept their new watches, preferring to sell the established Swiss imports. The

Pitkin watches were perhaps fifteen years ahead of their American successors.

The breakthrough came from Dennison, Howard and Curtis who, perhaps realising the superiority of the Swiss methods, sent Dennison to Europe to learn all he could. In 1850 they founded the American Horologe Company in Roxbury: successively it became the Warren Manufacturing Company, the American Watch Company in 1859, the American Waltham Watch Company in 1885 and, in 1906, the Waltham Watch Company, which continued in business until 1950 – a 100-year span. From the beginning their products were lever watches wound by key, button wind being introduced in 1870. Eventually they had no fewer than 30 factories. The Waltham company produced watches of various qualities, and they were joined in this field by Howard in 1857, Elgin in 1864 (Elgin were called the National Watch Company until 1874) and the Hamilton Company in 1892.

A peculiarity of all American watch companies is the variety of names used on dials and movements. The personalised names came from partners of firms, and the changes of company names from amalgamations, financial problems, changes of site and so on. There was also a tendency to use monograms instead of names on dials.

The better-quality American watches of this period were lever watches and they often had quite elaborate methods of regulation control, with fine screw adjusters of various designs, and attractive engine engraving on the movements. The cases were double bottomed in the European style for key wind, but there was also an interesting style with a screw-on front and button wind using rocking-bar hand adjustment – to allow the movement to be hinged from the case to give access to the regulator, the winding stem was partially withdrawn by pulling the button.

An alternative line of development was followed by the Waterbury Watch Company. Founded in 1878 by Benedict and Burnham, the company was called Waterbury from 1880 to 1898, the New England Watch Company from 1898 to 1912, and in 1914 it was sold to Ingersoll, who finally failed in 1922. At the beginning in 1878, Buck obtained a patent for what was to become the Waterbury Watch. It was a cheap, machine-made design with only 58 parts and a very long mainspring which was

American watches. *Top left and right* This is a watch with a Swiss lever escapement by the Peoria Watch Co, Illinois. It has a compensated balance, a going barrel and fine adjustment of the regulator by worm and wheel. The case is made of silver with a screw-on front. The hands are adjusted by a rocking bar operated by a lever hidden under the bezel. *Bottom left* This watch by the American Waltham Watch Co has a Swiss lever escapement, a compensated balance, a going barrel and fine adjustment of the regulator. The gold-plated case has a screw-off back. *Bottom right* This watch is by the Waterbury Watch Co and is a late example of their duplex watch. It does not have the long mainspring and rotating movement. The pressed-out escape wheel is shown separately. The dial is made of paper and the watch case of plated steel. The hands are set by pushing and rotating the winding button.

infamous for the time it took to wind. This spring was coiled
behind the movement which rotated like a tourbillon, carrying
the minute hand around with it. The dial was printed on paper
and covered in celluloid, and the duplex escapement had a
pressed-out wheel. This remarkable watch was discontinued in
1891, although the duplex escapement continued to be used
with conventional winding until the end of the century. Some
jewelled models were also produced. The company failed in 1914
and was then acquired by the Ingersoll brothers, who had started
their pin-lever watch business in 1892 using a small clock movement complete with a folding winding key inside the case back.

Left A pin-lever watch by F. E. Roskopf with a silver-plated cast-metal case of about
1910. The casting on the back of the case is of a railway engine, and the bezel is surrounded by the words 'Veritable montre chemin de fer'. The watch has a going barrel. *Right* A Roskopf movement from the late nineteenth century which also has a
going barrel. The escape wheel and lever can be clearly seen.

Thus it can be seen that in America during the period
1850–1900 the quest for volume production at a price and quality to compete with Swiss products led to the formation of a large
number of watch companies. Many of these failed to survive and
were sold, combined or taken over. There was no signing of
movements by individual makers as there was in Europe, since
from the very start the concept was one of the machine-made
factory product. This devotion to production had its effects in

Europe, for after the challenge of this approach was realised by the Swiss, they amended their own methods and eventually eliminated most European competition. The cheap watch had developed independently in Switzerland in the form which was eventually to become worldwide. Roskopf designed and produced a pin-pallet lever watch in 1867. He presented one of his watches at the Paris exhibition and the award of a medal contributed to his success. Roskopf used the minimum number of parts – a three-wheel train, a simple, keyless winding with hand adjustment by finger, and cheap metal (as opposed to gold or silver) cases. His watch was patented abroad but was available to Swiss makers, who were able to produce similar designs. Thus Switzerland began to present large numbers of pin-pallet, cylinder and lever watches which together satisfied all price levels of the world market. The pattern of watchmaking for the first half of the twentieth century was set.

Time problems

Although the development of the chronometer in the late eighteenth century had solved the problem of navigation at sea, there were still problems on land. Watch hands are constrained to rotate at a constant speed, and in any particular locality a watch or clock could be checked with a sundial. Unfortunately the sun does not behave in the same way as a watch, and at times it is 'ahead' of the watch and at others 'behind'. The 'equation of time' is the difference between 'solar time' and 'mean time', and any person adjusting his watch by use of a sundial needed to know the value of the equation of time. These values were available, sometimes printed on 'watch papers' kept in the back of a pocket watch. In a community, the most important timekeeper to be kept correct was the church clock, which anyone could then use to adjust his watch. This was acceptable when long-distance travel was uncommon.

The advent of the railway was significant because the timetable needed to be set in a standard time to avoid accidents. In the UK, the 1880 Definition of Time Act established Greenwich time, which was to be kept over the whole country. The aims of the Act were achieved in practice by the use of the Post Office telegraph system. The problem became world wide as overseas travel became more common, and in 1884 the international time zone was established. Later, radio time signals solved such problems. One further difficulty was caused by the latitude of various countries, and some local adjustments of clocks and watches had to be made, sometimes known as 'daylight saving'.

1900 ONWARDS

From 1900 onwards there are few new technical developments. Probably the most significant was work in the metallurgical field. A change in temperature causes changes in the dimensions and elastic properties of materials, which called for the application of temperature compensation: this was done in the form of the cut, bimetallic balance. Although this design of balance gives an enormous improvement, it still leaves a middle-temperature error between, and errors beyond, the two temperatures at which the balance can be designed to give correct compensation. These errors may be given an allowance based on the rate of gain or loss; the allowance will vary with ambient temperatures.

The early approaches to eliminating this error were by means of auxiliary compensation devices, in which the inward or outward motion of the balance arms due to change in temperature was modified by extra fittings. Many of these compensations were discontinuous in action and only partially corrected the error, and even the best continuous-action designs were not perfect at all temperatures. In 1900, metallurgical investigations into the properties of nickel-iron alloys enabled Guillaume to produce an alloy which was used in the compensated balance in place of steel. This alloy has a temperature coefficient of expansion of such a value that when it is combined with brass in a cut-compensated balance (known as a Guillaume balance) the middle-temperature error is virtually eliminated. Guillaume had previously produced an alloy called Invar which had a negligible temperature coefficient of expansion, and he continued his work to tackle in a better way the errors caused by variation in elastic properties. In 1912 he deduced that it was possible to make an alloy for balance springs which would have no change in elasticity with temperature. Experimental work to produce the alloy Elinvar proceeded, and from 1919 it was possible to use a monometallic balance of Invar controlled by an Elinvar spring. These alloys were not ideal, because their other properties, such as hardness and suitability for being formed, were not as good as those of the metals they were replacing; but following Guillaume's lead, modern metallurgy has produced more suitable materials, some of which also have the desirable properties of being non-rusting and non-magnetic.

From pocket watch to wrist watch

Changes in the non-technical field were dominated by the switch from the pocket watch to the wrist watch. A wrist watch had appeared in 1571 and in 1809, but it was not until the early years of the twentieth century that any quantity was produced. The tendency to change was enhanced by the 1914–18 war, in which a watch on the wrist proved much more convenient than one in the pocket, and when the new design had a chance to prove its reliability. From the manufacturer's point of view, the first problem was one of scale. However, the volume-produced, Swiss cylinder fob watch used movements of such a size to show that the problem was not insoluble, and early wrist watches made in Switzerland used the cylinder movement. The Swiss lever wrist watch followed some years later.

By 1930 the ratio of wrist watches to pocket watches produced was of the order 50 to 1. Winding was by button and hand adjustment by rocking bar or shifting sleeve. The dial was arranged so that the winding button was at 3 o'clock; however, watches can be found with the button at 12. These may be fob conversions by a local watchmaker to satisfy a customer or to produce an up-to-date line to sell. The strap lugs in early designs (and conversions) were simple wire loops added to what was virtually a miniature pocket watch case or fob case without decoration. Hinged or snap bezels and backs were used, the snap design giving better access to the movement, which was removed from the front of the case. Dials were white enamel or metal and without decoration, but the numbers were sometimes treated with a luminous compound. Watch glasses began to be made from a transparent plastic material, less fragile than glass but which scratched more easily and also tended to become yellowed with age.

Although the pocket watch continued until the end of the 1939–45 war, production after 1945 was minimal. The demise of the waistcoat as a common form of dress was a contributory factor to its decline. The change from the pocket watch, coupled with the economic effects of large-scale Continental production and the depression, effectively halted the production of watches in England by 1930, and it was not until after the end of the 1939–45 war that a new machine-based industry was established. In the early years of the century English watchmakers had started to use the going barrel, though some continued to use key wind. There were also a number of firms selling watches with their name on the dial or plate – but also somewhere in small letters were the words telling the story: 'Swiss made'. This practice continued throughout the period.

The cylinder movement was another casualty at this time. Having demonstrated the viability of the wrist watch, the development of the miniature lever movement meant that there were three types of movement looking for room in a market which really only needed two. For the expensive fully or partly jewelled watch the lever was ideal, and for the cheap watch the pin pallet was satisfactory. Thus, after 200 years the cylinder movement had completed its useful life.

In 1945, the European watch industry was dominated by Swiss-produced wrist watches with lever movements. After this date the quality wrist watch began to acquire the complexities that had been available in pocket watches. Automatic winding, using Perrelet's idea of a swinging weight which was conceived in c.1775, was often used. This was not the first modern application of automatic winding, for Harwood had used Perrelet's ideas in 1924 and made wrist watches in the Isle of Man. Harwood had completely dispensed with the stem winder with shifting-sleeve hand adjustment, but the later Swiss designs kept the winder/hand adjuster in the majority of cases. One of the problems in the development of automatic wrist watches is the excess of movement available at the wrist. The chronograph also became available in wrist form, often coupled with datework, alarm work, moonwork etc. Watches became shock resistant, waterproof, acceleration proof, able to function on land or sea, in the air, under the water, in space under weightless conditions and on the moon.

Despite the Swiss domination, there are watch industries in most of the major countries in Europe, but with standardisation and the operation of multi-national companies it is not easy to distinguish national characteristics in the products of any one country. The pattern of development in America was similar to that in Europe. The need for pocket watches slowly disappeared as the wrist watch became established. American companies continued to proliferate and to succeed or fail as in the years before the turn of the century. The depression stabilised the situation, so that the post-war industry is based on fewer large companies. The history of these companies is complex and it is not easy to trace all the changes. The pattern of production became similar to that in Europe, with the lever escapement for the more expensive watch backed by the pin pallet for the cheaper range.

Electrical developments

New developments in America were concerned with electrical applications and it seemed that the new designs would see the end of mechanical watches. The first examples of the new breed of watch in 1957–8 used mechanical switching of the battery power supply to impulse the balance wheel at the appropriate instant; but in reality these watches were just another form of 'automatic' watch which needs no winding.

Later designs of watch, based on the development of the transistor and integrated circuit technology, did lead to an entirely different form of watch. The developed 'electronic' watch has a quartz crystal vibration rate of 32,768 per second (32.768kHz) rather than the scant 2.5 vibrations per second of the Victorian pocket watch. This factor makes for excellent timekeeping at a spectrum of prices.

The range starts with the cheap, plastic-cased, digital-display watch which usually offers 12- or 24-hour-based time together with day and date as the normal display. If the rather fragile push buttons are used, the watch may also be used for hourly signal, alarm-setting, split-second chronograph and to illuminate the display in the dark. Similar analogue-display electronic watches are also available at moderate prices. The development of these watches will be outlined in the next chapter.

As suggested above, it seemed that the new watch would destroy the market for mechanical watches; however, this has not been the case. Indeed, during the 1990s, after an initial interest

in the new technology, the demand for mechanical watches,
especially automatic, is now re-established.

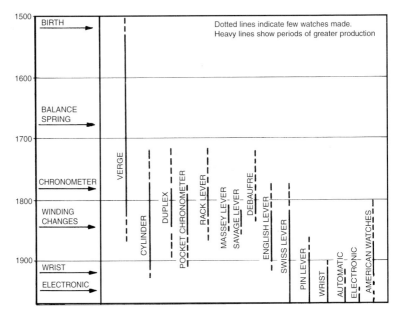

Fig 11 Chronological chart.

Fig 11 shows the lifespan of the watch escapements discussed
and some of the significant happenings in the four-and-a-half
centuries reviewed.

CHAPTER 2

TECHNICAL SURVEY

A watch is designed to indicate time by the rotation of hands against a calibrated dial. The power causing the hands to rotate is obtained from the mainspring coiled in the barrel. The power is transmitted through a (gear) train which allows the hands to rotate at a different speed to the barrel, the hour hand going more slowly and the minute hand faster. In a gear train, the ratio of the rates of rotation of two meshing wheels is the inverse of the ratio of the numbers of teeth. Fig 1 shows the arrangement of a watch with a normal four-wheel train.

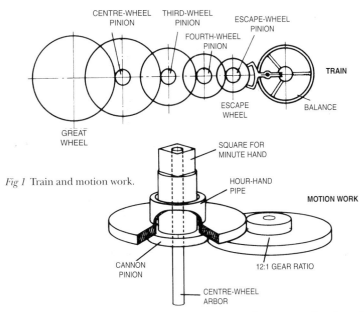

Fig 1 Train and motion work.

The first wheel in the train is known as the great wheel and is either the fusee wheel, or, in a watch with a going barrel, the barrel wheel, or an extra 'dummy' wheel inserted in the train to make keywinding anticlockwise as in a fusee watch. The great wheel meshes with the centre-wheel pinion, which is on the arbor or shaft of the centre or second wheel. On the end of the same wheel shaft is the cannon pinion to which the minute hand is attached. The cannon pinion is a good, but not rigid, fit on the centre-wheel shaft so that the hand may be set without moving the train. The centre wheel must be constrained to rotate once per hour. The hour hand is driven concentrically with the minute hand through a separate twelve-to-one step-down gear train known as the motion work. The motion work is situated between the dial and the dial plate, whereas the train wheels are situated between the plates. Continuing with the train layout, the centre wheel meshes with the third-wheel pinion, which is on the same shaft as the third wheel, which in turn meshes with the fourth-wheel pinion, which is on the same shaft as the fourth wheel. This fourth wheel has the seconds hand fitted to its shaft and must therefore be constrained to rotate once per minute. These four wheels have their numbers of gear teeth chosen so that the seconds hand and minute hand rotate at the correct relative speeds, but if the mainspring is wound the train would rotate at enormous speed. The speed is regulated to the correct value by the rate of vibration of the balance wheel through the escapement.

ESCAPEMENTS

The action of an escapement allows the escape wheel to rotate one tooth at a time. The escape-wheel shaft carries a pinion which meshes with the fourth wheel of the train, so that the train is allowed to rotate in small increments which give the correct average speed. The different designs of escapement may be broadly divided into two groups: frictional rest escapements and detached escapements. Examples from either group may have recoil or dead-beat characteristics, and there are also examples from either group known as single-beat escapements, which allow only one escape-wheel tooth to pass or escape per balance vibration, compared with the more usual two per vibration. In the frictional rest escapement, the vibration of the balance is never free from friction forces owing to contact with the escapement; examples are the verge and cylinder escapements. In the detached escapement, the vibration is free of the train except at the instants of unlocking the train and receiving impulse to

maintain the vibration. The detached escapement has considerable advantages in timekeeping qualities; examples are the detent
and lever escapements. Recoil characteristics are exhibited by
designs in which the escapement, and hence the train, moves backwards for an instant during the escapement action; examples are
the verge escapement and the lever escapement with draw. An
escapement which is dead beat does not have this recoil action, the
cylinder escapement being an example.

We will now consider a number of different escapements in
detail. Not all these designs were made in the same numbers, and
the duplex, detent, Debaufre, rack lever, Massey, lever and
Savage lever are less common than the rest. (Some uncommon
escapements are not described.)

Fig 2 Verge escapement.

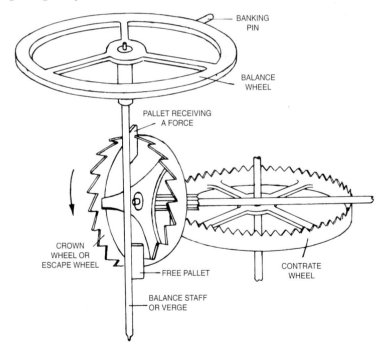

Verge escapement

Fig 2 shows the verge escapement, used from the sixteenth century to the nineteenth century. It is a frictional rest, recoil

escapement. Before its use in the watch it was used in clocks, but its origins are not clear. The escape wheel is sometimes called the crown wheel because of the shape of the teeth. It is arranged to engage with two pallets on the balance staff or verge. The shaft of the crown wheel is at right-angles to the remainder of the train so that the fourth wheel, with which the crown-wheel pinion meshes, is a contrate wheel, with teeth in the vertical plane of the staff, rather than a conventional gear wheel.

The crown wheel has an odd number of teeth (usually 13), and as it rotates the pallet at the upper end of the verge is pushed by one of these teeth, causing the balance to rotate until the tooth escapes past the end of the pallet. The train and balance are momentarily free, but the rotation of the verge has been just enough to cause the lower pallet to engage with a crown-wheel tooth at the opposite end of the diameter (the pallets are a bit more than a right-angle apart), and the freedom ceases. The rotation of the balance is stopped by this lower pallet but not instantaneously, and there is a small arc of motion in which the crown wheel and train rotate backwards or recoil. The influence of the mainspring causes the crown wheel to resume its normal direction of rotation and to push the lower pallet out of its path, until the tooth escapes and the sequence is restarted. The crown-wheel teeth are undercut to ease the recoil action. If the pallet does not succeed in bringing the balance to rest, banking will occur, when a small pin in the balance rim strikes a projection on the balance cock (upper bearing), thus avoiding excessive swing.

Although the balance and verge are vibrating and continually reversing their direction of motion, the fact that the escaping teeth are at opposite ends of a crown-wheel diameter means that the rotation of the crown wheel and the train is undirectional.

Cylinder escapement

The cylinder escapement was introduced in 1726 by Graham. This may have been as a result of his earlier association with Tompion, Booth (or Barlow) and Houghton, who had a patent in 1695 for a new escapement which could be considered a forerunner of the Graham design. The cylinder escapement was also called the horizontal escapement, so as to distinguish it from the vertical or verge escapement.

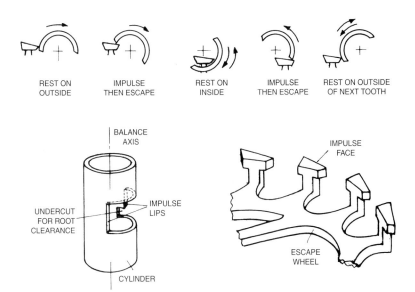

Fig 3 Cylinder escapement.

The escape wheel has a vertical shaft and usually 15 teeth which project vertically from the rim of the wheel. These teeth engage with, and pass through, a special slot cut in a hollow cylinder fitted into the balance-wheel shaft. The working portion of the hollow cylinder has approximately half its circumference removed, so that a tooth may escape by passing through the cut-away portion of the cylinder at the appropriate moment during the balance vibration, but will rest on the outside or the inside of the cylinder during the remainder of the vibration. The retained half of the cylinder is undercut to allow the tooth root clearance when it rests on the inner surface of the cylinder.

The escapement is a frictional-rest, dead-beat design in which impulse is given to the balance twice per vibration, by the sloping surface of the escape-wheel tooth acting on the edge of the hollow cylinder. Fig 3 shows the sequence of events in the action of the escapement.

Duplex Escapement

The duplex escapement derives its name from the original double escape wheel used in the design. The more common form, described here, has two sets of teeth on a single escape-wheel rim.

One set of teeth is for locking the train and the other for impuls-
ing the balance. The teeth are set alternately around the rim.
Impulse is given once per balance vibration, since the locking
teeth only escape when the balance is rotating in the opposite
direction to that of the escape wheel. The impulse is given by the
inner ring of raised teeth on the rim of the wheel striking the
impulse pallet, which projects from the balance shaft or staff.
The outer, pointed teeth projecting from the escape-wheel rim
are allowed to escape by passing through a vertical slot cut in a
jewelled roller fitted around the balance staff.

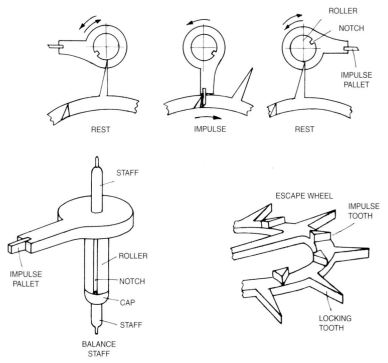

Fig 4 Duplex escapement.

Escape is in two steps: a very small movement as the tooth slips
across the width of the slot, and a much larger movement as the
tooth escapes from the slot. This motion is stopped as the next
pointed tooth comes to rest on the roller surface. On the return
swing of the balance, the slot slips past the tooth resting on the
roller without permitting it to escape. This tooth is pressing on

the roller for almost the complete vibration, and the escapement is frictional rest with single-beat action. Fig 4 illustrates the action of the escapement.

Detent escapement

The word 'detent' is defined in the *Concise Oxford Dictionary* as a 'catch by removal of which machinery is set working'. This definition is almost adequate to describe the escapement used in the pocket chronometer and shown in Fig 5. The detent is sprung or pivoted about one end, and it holds the escape wheel and train locked by means of a jewel (known as the locking pallet) standing up from its surface. The balance staff carries two pallets and as the balance swings, the discharging pallet briefly pushes the passing spring against the detent, which moves to release the escape wheel. The rotation of the escape wheel then allows impulse to be given to the impulse pallet on the balance staff by an escape-wheel tooth, before the train is again locked by the locking pallet on the detent. On the return swing of the balance, the passing spring merely flexes as the discharging pallet passes, and does not unlock the escape wheel and train. Thus escape

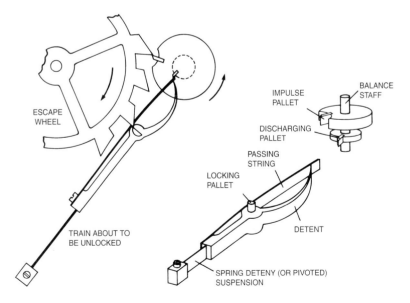

Fig 5 Detent escapement.

only occurs once per vibration. During the vibration, the balance is detached except at the instant of unlocking and impulse. The escapement is a detached, dead-beat design with single-beat operation.

Early pocket chronometers designed by Arnold had a pivoted detent but later he used a spring-detent design. Earnshaw also used a spring-detent design. Most of the other English pocket chronometers seen will also have spring-detent escapements but Continental makers usually favour the pivoted detent. The relative merits of the designs are quite difficult to assess, but one advantage that the spring detent offers over the pivoted detent is that the escapement requires no lubrication except for the balance and escape-wheel pivots.

Debaufre-type escapements

These escapements, like the verge escapement, use a contrate wheel as the fourth wheel and at first sight might be taken as a

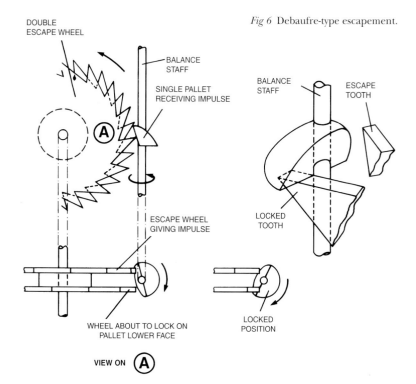

Fig 6 Debaufre-type escapement.

verge design. Closer inspection will reveal that the simple crown wheel and verge has been replaced by other designs. Various arrangements have been used, involving one or two pallets attached to the balance staff with an escape-wheel system comprised of a single or double crown wheel, or a single or double saw-tooth wheel. Fig 6 illustrates one variety. The basic idea was conceived by Debaufre in 1704, but the main use occurred at the end of the eighteenth century and the beginning of the nineteenth century in England. The escapements are frictional rest and have a variety of names such as dead-beat verge, club-foot verge, Ormskirk and chaffcutter.

The pallets are horizontal, D-shaped and have inclined edges. They alternately receive impulse from an escape-wheel tooth on the inclined edge, and lock the train on the horizontal surface. If two escape wheels are used, the single pallet receives the impulse as one wheel escapes and then locks the other wheel. If two pallets are used, the single escape wheel gives impulse to the one pallet before being locked by the other pallet.

Lever escapements

The basic principle of the lever escapement is shown in Fig 7. The lever is pivoted at A. Impulse is given to the pallet B by the escaping tooth C. This impulse is transmitted to the balance. The

Fig 7 Principle of the lever escapement.

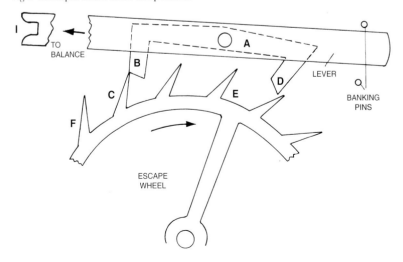

rotary motion of the lever created by the impulse causes pallet D to intercept tooth E and lock the train. On the return swing of the balance, the lever is pivoted back to allow tooth E to escape as it gives impulse to pallet D, which is again transmitted to the balance. The lever motion causes pallet B to intercept tooth F to lock the train. This continuous rocking motion of the lever allows teeth to escape successively. There are a number of varieties of the design. Those discussed here are mainly concerned with the connection of the lever to the balance staff, but two have differences from the pallet and escape-wheel design shown in Fig. 7.

The original lever designs did not have draw, which is essential in all detached forms of the escapement to avoid accidental motion of the lever, which would result in friction between the lever safety piece and the balance-staff roller. The friction would disturb the timekeeping quality of the escapement. Draw consists of shaping the escape-wheel teeth and lever pallets so that the lever is pulled towards the escape wheel. The lever cannot be drawn in too far because banking pins are placed to restrict the angular movement. When the lever is deliberately moved by the return swing of the balance, draw causes a slight recoil in the action. Draw is probably absent if the locking faces of the lever pallets have convex curvature, but the use of straight-faced pallets will not guarantee the presence of draw. However, it is unlikely that any lever made after 1825 will lack draw.

In the form shown in Fig 7, the escape wheel has pointed teeth and the actions of lifting and pivoting the lever are wholly a result of the slope on the pallets. This is characteristic of the English forms of the lever escapement discussed (the rack, Massey, Savage and table-roller levers), whereas in the Swiss lever the shape of the escape-wheel teeth will be seen to be modified to share the lifting action between pallet and teeth, and in the pin lever the escape-wheel teeth are steeply sloped to give most of the lift.

The lever escapement descriptions that follow are not exhaustive of the types that may be found. However, although there are other rare types, the majority of the variations seen will only be trivial. For example, the English layout may be used with the Swiss-type escape wheel which shares the lift between lever pallet and escape-wheel tooth. Similarly, levers may have different shapes with counter-balances and suchlike, but the action and acting surfaces are the same. A variation which is of more interest but would require careful measurements is in the proportions of lever lengths, roller diameters etc, which will affect the forces involved in the action of the escapement and the time taken for each part of the action to occur.

Rack lever escapement

The rack lever design was conceived by Abbé de Hautefeuille in 1722, but the particular form discussed here was patented by Litherland in 1791. It was made in England in some numbers in the late eighteenth and early nineteenth centuries. The operation of the escapement is essentially that described above and shown in Fig 7 on page 57, and the lever is connected to the balance by a rack which is permanently engaged with a pinion on the balance staff. Thus the escapement is not detached, nor is it really a frictional 'rest' design.

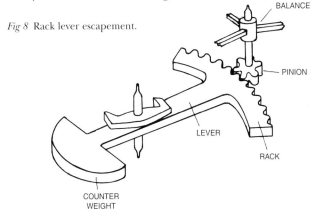

Fig 8 Rack lever escapement.

The large mass of the rack at one end of the lever is counter-balanced by a D-shaped piece at the other end. Many rack lever escape wheels are of large diameter, with 30 teeth instead of the more usual 15. This large wheel is associated with a three-wheel train and is therefore not suited to the use of a seconds hand in the conventional way. The three-wheel train will also mean that the direction of rotation of the escape wheel is opposite to that of the more usual 15-tooth escape wheel. Fig 8 shows the lever and balance staff arrangement. It may be seen that the balance staff, lever shaft and escape-wheel shaft are planted so as to form a right-angle. This arrangement is characteristic of developed forms of the lever escapement in England.

Massey lever escapements

In 1814, Massey invented a detached lever with the right-angled layout mentioned above. It has been suggested that the original concept was developed from the rack lever, by forming a single-toothed 'pinion' on the balance staff to engage with a two-tooth 'rack' on the lever end. This original form is shown in Fig 9, and in it the roller on the balance staff is entirely made of steel, which is in harmony with the idea of the single-tooth pinion. Some of these early original-form Massey lever watches also used the large-diameter, 30-tooth escape wheel with a three-wheel train. Draw is not found in early Massey levers, but was introduced and is usually present in the later forms.

Fig 9 Massey lever escapement.

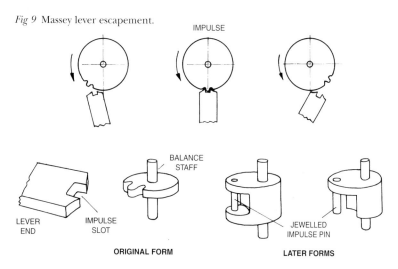

In these subsequent developments, the single tooth of the pinion or roller became a jewel forming a crank to engage with the lever slot, and the escapement is sometimes called a crank lever or crank roller escapement. In this form (see Fig 9, later forms), the significance of the Massey escapement as a lead toward the final form of the English table-roller lever escapement is clear.

Although Mudge and others had made lever watches between 1770 and 1800, the numbers were minimal and the escapement did not seem to be making any headway as an alternative to the verge or cylinder. Massey's design, which survives in some quantity, especially in the later forms, really set the lever on the path to success. This may have been a result of a combination of circumstances rather than the design, but the dominance of the lever could be considered to start in 1814.

The action is similar in all forms. As the balance swings through the centre arc, the roller tooth or jewel engages with the slot in the end of the lever and unlocks the escape wheel. The jewel then receives impulse from the slot (transmitted from the escape wheel) and the balance becomes detached for the remainder of the action. The escapement is locked by one of the pallets on the lever during the detached portion of the vibration. On the return swing, the roller tooth or jewel again picks up the lever slot, unlocks the train and receives impulse. While the balance is in the detached portion of the vibration, the lever should remain drawn in to the escape wheel. Should it move from this position because of the absence of draw or the failure of the draw action, then the curved surfaces of the outer sides of the lever's slotted end (lever horns) will rest on the cylindrical surface of the balance roller, preventing the train from unlocking and damaging the escapement. This safety action will introduce friction and destroy the advantage of the detached balance in timekeeping ability, since for the faulty part of the action it is frictional rest in character. When the return swing of the balance brings the roller tooth or jewel back to the appropriate position, the engagement with the lever slot will restore correct action so that the friction is only a temporary feature.

Savage two-pin lever escapement

The Savage two-pin design is another detached lever escapement invented in about 1815. It uses the right-angled layout, but the arrangement for locking and unlocking the train and for impulse is different to other forms of lever. As shown in Fig 10 overleaf, the balance staff is fitted with a roller which has two pins protruding from its surface. Situated centrally between these pins is a slot. The lever end has a wide opening with a jewelled pin just

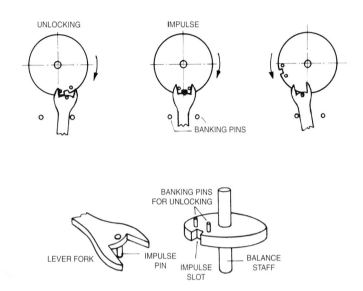

Fig 10 Savage two-pin escapement.

below the centre point of the bottom edge of the opening. As the balance swings, one of the pins on the roller enters the jaws of the wide lever opening. This unlocks the train, and impulse is transmitted from the escape wheel to the balance roller as the pin on the lever engages with the slot in the roller. The train is then locked by one of the lever pallets in the usual way. The balance becomes detached for the remainder of the vibration except for the repetition of the action on the return swing, when the second pin on the roller effects the unlocking of the train. Should the lever become displaced during the detached portion of the vibration, the lever jewel will rest on the side of the balance roller giving a safety action with frictional rest until the normal unlocking and impulse restore the situation.

English table-roller lever escapement

The table-roller lever is the final form taken by the detached lever escapement in England. It first appeared in about 1823 and had a lifespan of approximately one hundred years. It is shown in Fig 11. The right-angled layout is again used and the escape-wheel teeth are pointed. The slot end of the lever is a different shape to the Massey and Savage forms. The roller on the balance staff has a single elliptical or D-shaped jewel protruding from its surface. As the balance vibrates through the central position, the roller jewel comes into contact with the lever and engages the slot. The train is unlocked and impulse is transmitted from the escape wheel via the lever to the roller jewel. The balance then swings detached from the lever and train until the action is repeated on the return swing.

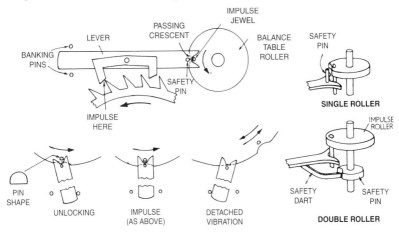

Fig 11 English table-roller lever escapement.

Should the lever become displaced from the correct position when the balance is in the detached portion of the vibration, the safety pin situated at the bottom of the lever slot will come into contact with, and rest upon, the outer surface of the balance roller. The frictional-rest safety action will continue until the roller jewel engages with the lever slot. The outer surface of the roller has a small passing crescent in its surface to allow the safety pin to pass during the normal unlocking and impulse action.

As an alternative to the safety pin and passing crescent on the single-roller arrangement, two rollers may be used. The first, called the impulse roller, has the impulse jewel fitted and performs the escapement action. The second, smaller-diameter safety roller with the passing crescent operates with a safety dart protruding from the underside of the lever rather than a safety pin. This is also shown in Fig 11.

From 1850 onwards almost all the watches made in England had the detached table-roller lever escapement. In this form the escapement is long lasting and accurate, and many watches initially fitted with other escapements were converted to detached lever late in their working lives.

Swiss lever escapement

Whereas the English lever escapement has been shown with the balance, lever and escape wheel pivots forming a right-angle, the Swiss lever design is often arranged with these components in a straight line (see Fig 12). The lifting and pivoting of the lever is caused partly by the shape of the lever pallets and partly by the shape of the escape-wheel teeth. These teeth have a sloping surface on a broad foot rather than coming to a point. The purpose of the change in shape is to divide the lift between lever and escape wheel, which is now said to have club-tooth form.

The action of the escapement is the same as that described for the English lever. The Swiss form frequently has the double-roller arrangement with a safety dart protruding from the lower surface of the lever. Thus the swing of the balance allows the impulse

Fig 12 Swiss lever escapement.

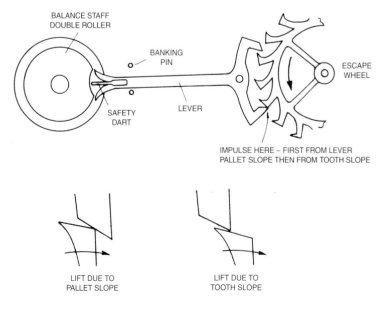

jewel on the impulse roller to engage the lever slot, unlock the train and receive impulse. The balance then becomes detached while the lever is held to the banking pins by draw. Safety action by the dart resting on the safety roller will occur should the lever move from the correct position.

The Swiss lever escapement is now the only mechanical escapement made in large numbers for high-quality watches.

Pin-lever escapement

The pin-lever escapement, developed from the Roskopf design of 1867, is a cheap escapement for mass production. The original concept of producing a watch for the ordinary man has been overtaken by the mass production of jewelled Swiss lever escapements, but the pin-lever escapement is still the normal arrangement of the very cheap watch; it is the only alternative available to the Swiss lever (excluding electronic designs, which do not use a balance-wheel arrangement).

Fig 13 Pin-lever escapement.

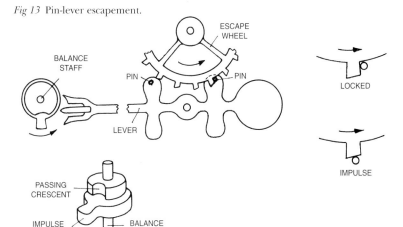

In the pin-lever design, the impulse pallets on the lever are replaced by steel pins. These pins perform the same duty as the jewelled pallets, in that they allow the escape wheel to rotate in increments as they alternately unlock the train to receive impulse and lock it. The escape-wheel teeth supply most of the lift and have steeply sloped faces. This may be seen in Fig 13. At the end of the lever the impulse is transmitted to the balance wheel by a forked end with a wide slot, which engages with a metal impulse piece on the balance staff. Safety is by means of a dart operating on a roller, with a passing crescent fitted to the balance staff. An alternative design uses a pin on the balance roller or balance arm to receive impulse.

The Roskopf design used a right-angled layout, but pin-lever escapements are also made in straight-line form. Early pin-lever watches used a three-wheel train. However, if a seconds hand was required, a fourth wheel was added to the train to allow conventional 'clockwise' rotation of this extra feature. Most Timex mechanical watches use a modified design of pin-lever escapement (US Patent 3,146,581, 1 September 1964 to G. F. W. Garbe).

TIMEKEEPING

A consideration of timekeeping is now in order. In early watches timekeeping was poor, but the application of the balance spring gave such an improvement that the disadvantages of frictional-rest escapements became obvious and a variety of other escapements was developed. The influence of the desire to navigate successfully has been mentioned on page 22, and it was the spur to many makers.

The knowledge that a detached escapement was essential in order to achieve accurate timekeeping with ease (although Harrison demonstrated accuracy with a frictional-rest design) led finally to the detent and lever escapements. All good watches had one of these by 1850. The frictional-rest cylinder escapement lasted much longer, mainly for a market dominated by price. However, it finally succumbed in the late 1930s, leaving the lever watch as the only choice, either in jewelled Swiss lever form or in the cheaper Roskopf design. This is still the case today, except that the electronic watch has also taken a share of the market.

Temperature effects

Accepting the detached escapement as a necessity, the problems to be met to achieve good timekeeping were dominated by the effects of temperature. Temperature change affects timekeeping in three ways. First, and most important, the balance-spring metal is less springy at high temperatures and more springy at

low temperatures: thus a watch will lose at higher temperatures
and gain at lower temperatures. Secondly, temperature variation
causes the balance to expand or contract. A hotter balance that
has expanded has its mass at a greater radius, and under the
application of a constant torque will take longer to vibrate, so
that a watch will again tend to lose at higher temperatures.
Finally, expansion and contraction will affect the dimensions of
the balance spring. Some form of temperature compensation is
therefore required.

Initial attempts at temperature compensation used bimetallic
curbs, made of two thin layers – one of brass and one of steel.
When heated or cooled, the different rates of expansion of the
two metals will cause the composite material to bend. One end of
a curb is fixed, while the other, the free end, is arranged so that
it holds the balance spring close to its point of attachment to the
watch plate. If the watch gets hotter and tends to slow down, the
curb shortens the length of the balance spring and, it is hoped,
speeds the watch enough to compensate for the loss.

Curb compensation is an improvement over no compensation,
but interfering with the balance spring is not an ideal way of tackling the problem. The shape of the end curves of a balance
spring influences its isochronism – its ability to take the same
time for a vibration irrespective of the arc of vibration; and so a
curb compensating for temperature changes is liable to introduce other errors.

Fig 14 Curb compensation and compensated balance.

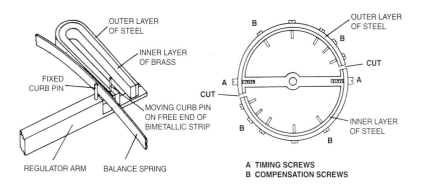

OUTER LAYER OF STEEL
INNER LAYER OF BRASS
FIXED CURB PIN
MOVING CURB PIN ON FREE END OF BIMETALLIC STRIP
REGULATOR ARM
BALANCE SPRING
CUT

B
OUTER LAYER OF STEEL
B
CUT
A
A
CUT
B
INNER LAYER OF STEEL
B

A TIMING SCREWS
B COMPENSATION SCREWS

A far better method was devised in which the balance rim was made of two different metals and then cut close to the balance arms (see Fig 14). The outer layer was made of brass and the inner layer of steel. When the balance temperature rises, the higher coefficient of expansion of brass will cause the outer layer to lengthen more than the inner layer, and the cut will allow the rim to bend inwards. Thus the balance mass will be moved inwards, and so the vibration time will be shorter. By carefully calculating the relative thickness of the brass and steel layers, the amount of compensation can be made appropriate to the loss expected due to the temperature rise. Similarly, if the watch becomes colder the balance rim will move outwards, compensating for the expected gain. The compensation obtained by this method is not exact, because the rate of variation of timekeeping due to change of elasticity with temperature is not always identical to the rate of variation of timekeeping due to change of balance dimensions with temperature.

Middle-temperature error

Elasticity and compensation may be examined mathematically, and it is suggested in Chapter 4 that the mathematically inclined collector might find some interest in thinking about timekeeping. A simple graph may be used to illustrate the effect of these different rates of change, and Fig 15 shows that it is possible for compensation with a cut-compensated balance to be correct at

Fig 15 Middle-temperature error.

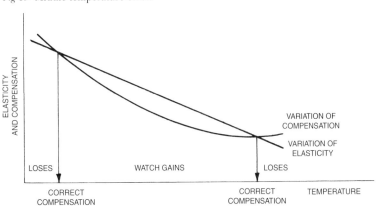

two temperatures where the two curves cross. But in between these two ideal situations is a small middle-temperature error, and in this area the watch will gain. Outside the two ideal situations the watch will lose. The effects of these errors can be minimised by manipulation of the mass and circumferential distribution of the compensation screws around the rim. This is a skilled task, however, for not only is the compensation affected by altering the mass, but so is the fundamental time of vibration of the watch. However, there are also two timing screws available at the ends of the arms which will not affect compensation but which can be altered to give a basic change in mass or mass distribution. Any alterations in a balance must also be made in such a way that its poise is maintained – that is, the centre of gravity is retained on the balance axis so that it does not have a 'heavy' side.

Middle-temperature error is a fact of life with the cut-compensated balance. There are numerous auxiliary compensations where additions to the balance are made in an attempt to limit excessive compensation, or to bring in extra mass when temperature changes are large, but no perfect solution has been found possible. These refinements are usually for use in the boxed marine chronometer and are rarely found in watches.

An alternative approach to middle-temperature error is a metallurgical one. In the late nineteenth and early twentieth centuries, special nickel-iron alloys were developed by Guillaume. One of these was made so that when used in place of steel as the inner layer in a cut-compensated balance, the compensation achieved was such that middle-temperature error was considerably reduced. Guillaume had also developed an alloy which did not expand or contract when subjected to changes in temperature. This alloy, known as Invar, has considerable possibilities for use as a balance material, for it would eliminate the second error mentioned above. However, the elasticity problems remained, and so Guillaume took his metallurgical work a stage further and worked out the correct alloying materials for a balance spring which would have no change of elasticity with temperature – and after some years of experiment Elinvar was successfully alloyed. Thus from 1919 it was possible to have a watch with Elinvar spring and Invar balance which should have minimal errors due to temperature change. These alloys did have some disadvantages, such as being prone to rust, difficult to work or susceptible to magnetic effects, but following the Guillaume lead, metallurgists have produced improved materials for use in watches. However, not all watches incorporate these advanced compensation techniques, for many watch users are completely satisfied with the performance of the non-chronometer watch, which is excellent.

Studies of balance-spring shape were made in the late eighteenth and early nineteenth centuries. A. L. Breguet produced the idea of an 'overcoil' at the outer end of the spring. In this design, the last coil of the spring was bent upwards to be above the other coils, and shaped to pass inwards over the lower coils of the spring and be pinned at a smaller radius. Breguet's ideas were studied by others, including Phillips and Lossier, and much later Billeter's book of Breguet balance-spring 'overcoil' shapes for various sizes of watch was produced (see Bibliography).

Vibration effects

Accuracy is affected by the rate of vibration of a watch. With a high-count train – that is, one with a high rate of vibration – one faulty vibration will have less effect than with a low-count train. The count of a train may be obtained by multiplying the numbers of teeth on the centre, third, fourth and escape wheels together and dividing by the product of the numbers of teeth on the third-, fourth- and escape-wheel pinions. This will give the number of complete vibrations per hour, since it is based on one revolution of the centre wheel. The count is usually twice the number of complete vibrations, since there are two escapes per vibration on all but duplex and chronometer watches. Bearing in mind that the fourth wheel must rotate once per minute if it is to indicate seconds, there is a considerable number of combinations of teeth on these wheels and pinions which may be used in a train. A common modern count is 18,000 per hour, which means that there are 5 escapes per second or 2.5 complete vibrations per second; older watches will exhibit lower counts – 14,400 per hour, representing a watch beating quarter seconds, is frequently used. The ultra-modern, quartz-crystal electronic watches with very high frequency vibrations have an inherently high capability for accuracy.

Positional error

As accuracy increases in a watch, errors which were relatively small begin to assume a new significance. One such error is a result of position. A clock or chronometer is arranged so it can be regulated to keep good time in one situation, but a pocket or wrist watch is required to give comparable accuracy regardless of its position. A series of tests on a good watch, in which temperature

Fig 16 Principle of revolving escapements.

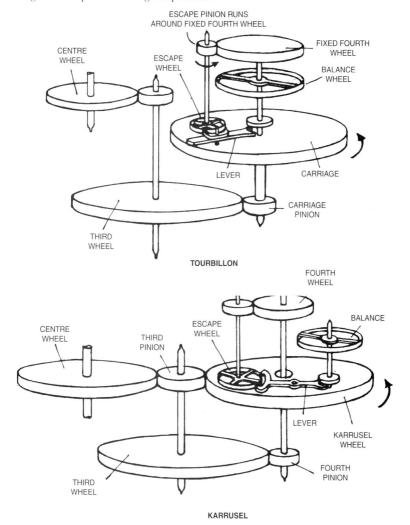

ESCAPE PINION RUNS
AROUND FIXED FOURTH WHEEL

CENTRE WHEEL

ESCAPE WHEEL

FIXED FOURTH WHEEL

BALANCE WHEEL

LEVER

CARRIAGE

CARRIAGE PINION

THIRD WHEEL

TOURBILLON

FOURTH WHEEL

CENTRE WHEEL

THIRD PINION

ESCAPE WHEEL

BALANCE

LEVER

KARRUSEL WHEEL

FOURTH PINION

THIRD WHEEL

KARRUSEL

effects have been minimised and for which the temperature effects remaining are to be allowed for by a rate of loss or gain, will show that this rate varies slightly, depending on whether the watch is tested with the 3, 6, 9 or 12 figure on the dial uppermost.

One method of dealing with positional error is to arrange that the escapement rotates steadily so that the error is averaged, and a single rate may then be used for the watch. This method is used in the tourbillon and the karrusel watches (see Fig 16). In the tourbillon, the escapement (balance, escape wheel etc) is mounted on a carriage which carries a pinion driven by the third wheel. The fourth wheel is fixed and is concentric with the carriage shaft. The escape-wheel pinion meshes with the fixed fourth wheel. Thus, as the carriage rotates, the escape-wheel pinion will roll around the fixed fourth wheel, which will cause the pinion to rotate and operate the escape wheel and balance in the normal way. The tourbillon was invented by Breguet in 1795, and in this design the mainspring power is transmitted to the escapement by the rotation of the carriage.

In the karrusel watch, patented nearly a century later by Bonniksen, the carriage is mounted on a karrusel wheel driven by the third-wheel pinion. The fourth-wheel staff passes through the centre of the karrusel bearing to allow the fourth-wheel pinion to mesh with the third wheel; power is transmitted to the escapement in the normal way, rather than through the carriage rotation as in the tourbillon. The karrusel rotates about once per hour, compared with the tourbillon which may rotate once per minute. Both these designs require considerable skill to manufacture, and are only found in watches of high quality.

EXTRA FUNCTIONS

A watch is often required to do more than indicate time. The most common complex mechanism indicates day, date, month and moon state. There are also chronograph watches which, as well as keeping normal time, have an extra mechanism to enable the watch to be used as a 'stop watch' without interfering with the normal display of time. There are watches which have repeating mechanisms which will, when activated, strike bell or gongs with a pattern that is recognised to give time to the last quarter-hour, half quarter-hour or minute.

Fig 17 Date work.

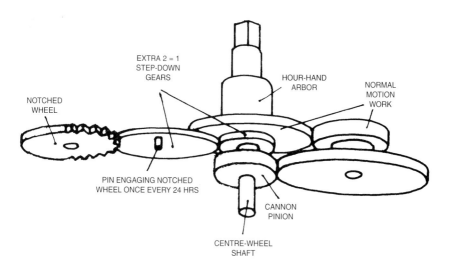

EXTRA 2 = 1
STEP-DOWN
GEARS

HOUR-HAND
ARBOR

NORMAL
MOTION
WORK

NOTCHED
WHEEL

PIN ENGAGING NOTCHED
WHEEL ONCE EVERY 24 HRS

CANNON
PINION

CENTRE-WHEEL
SHAFT

Date

Date, day, month and moon indication may be achieved by
extending the motion work. The centre wheel rotates once per
hour and the normal motion work is arranged to give a concentric rotation of the hour hand once every 12 hours. Day and date
need to advance once every 24 hours, so that a two-to-one step-
down gear from the hour-hand drive will give a wheel rotating at
the correct speed (see Fig 17).

One method is to fit this wheel with a single pin which is used
to move a notched wheel once per revolution. If a 31-notch
wheel carrying an indicator or pointer is used, the mechanism
will show dates but will need hand adjustment to allow for
months without 31 days. A similar arrangement may be used with
a 7-notch wheel to give days of the week. The month indication
may be operated externally by hand or by a dial driven by a single
pin from the 31-notch wheel. There must be an external adjuster
to allow days of the week to be phased with the date. Very complex
watches have been made with perpetual-calendar operation to
allow for both variation in the length of the month and for leap-
year variations. Moon indication from motion work is based on a
$29\frac{1}{2}$-day cycle using a 590- or 118-notch or tooth wheel.

Chronograph functions

The chronograph watch has three extra functions to perform: the starting, stopping and returning to the starting position of a seconds hand which is independent of the working of the watch. The starting and stopping is achieved quite simply by bringing a wheel (with extremely fine teeth or serrations) into mesh with another wheel, which rotates permanently with the main train of the watch. The engaging and disengaging is achieved by pushing the winding button or a special button which operates a spring-controlled lever system. The lever ends not concerned with engaging or disengaging feel their way into, and out of, slots in a column wheel, which rotates one step at each push of the button to control the stop, reset, start sequence. A third push on the button is used to reset the seconds hand. This is achieved by a special heart-shaped cam, which is pivoted in such a way that when the heel of a spring-loaded piece presses against it, the torque will bring the cam round to the starting position. It is held in this position with no overshoot by the foot of the spring-loaded piece applying the pressure.

The foot rests on a flat surface on the cam profile (see Fig 18, right). The profile is such that if the watch is stopped before the seconds hand has completed half a revolution, it will return in an anticlockwise direction, but if more than half a revolution has been made, the return is in the clockwise direction. The hand returns by the shortest route.

Fig 18 Chronograph reset mechanism.

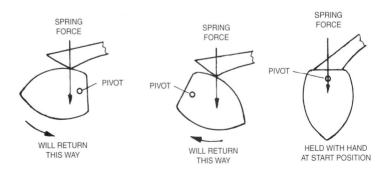

Repeating watches

The repeating watch is required to sense the time and then to make a sound on bell or gongs that conveys an indication of time to the listener. As an example, consider a quarter repeater which strikes on one gong, 'ting', to indicate the last complete hour, and on two gongs (one of which may be the hour gong), 'tang-ting', to indicate the last complete quarter-hour. Thus, 'ting, ting, ting, tang-ting' means that the time lies between 3.15 and 3.30.

The essential mechanism for producing this effect is quite straight forward, but the underdial work will look complex. The repeating mechanism has its own spring, which is wound by pushing the pendant or moving a slide piece on the watch case. The push action also operates a curved rack, which rotates a pinion attached to a wheel with 18 teeth. Twelve of these teeth are used to operate the hour strike-hammer, and two sets of three for the quarter-hour hammers. The amount of rotation given to this wheel must be such that the hammers indicate the correct time. From Fig 19 it may be seen that this rotation is controlled by a

Fig 19 Repeating mechanism.

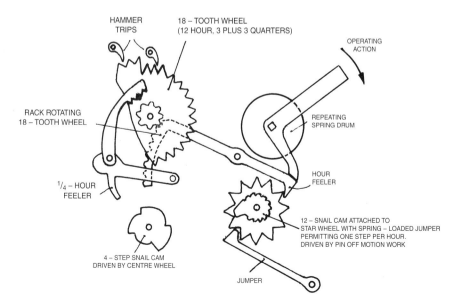

12-step snail cam (one step for each hour), which is driven by the motion work, and a four-step snail cam (one for each quarter) attached to the centre-wheel shaft. The amount of rotation is determined by two feelers which abut on to the two snail cams. Thus when the repeat mechanism is actuated, the spring is wound and the time is sensed by the position that the 12-tooth/3-tooth/3-tooth wheel takes up. When the pendant or slide is released, the spring drives the toothed wheel round, and each tooth trips the hammer as it passes it. For two sounds there are two trips. The speed with which the hours are sounded is controlled by the rate of rotation of a friction fly, which consists of a ratchet and spring-loaded pawl.

There are other designs of repeating watch which use similar mechanisms. Many incorporate an all-or-nothing piece which ensures that the correct time is sounded by not releasing the mechanism until the pendant or slide is fully depressed. If it were not included, a false time could be obtained by only partly pushing the pendant or slide. If this is done, the all-or-nothing piece allows the repeating spring to unwind but without sounding any gongs.

Electronic and electromechanical watches

Periodically in horological history there is an event which changes the perception of timekeeping. After the advent of the watch, which was made possible by the use of a coiled spring and balance 'wheel' rather than the weight and pendulum of a clock, these events might be considered to be: the application of the spiral balance spring; Harrison's marine chronometer; the invention of the detent escapement coupled with temperature compensation; the first lever watch; and the appearance of the electronic watch. The technology of the last event is entirely different to that of other watches.

An electric clock was developed and patented by Alexander Bain in 1841 (UK Patent 8783 of 1841), but it was not until the 1920s that such clocks achieved sufficient accuracy for observatory use. Quartz clocks were the next development, which used the vibration of a quartz crystal sustained by electric thermionic valve circuits. These circuits are bulky and demand considerable power: making a watch required much smaller components. In

practice, this means that the portable power source needed to be
about the same size as the spring barrel, and the quartz crystal
oscillator needed miniature transistor circuits. Accuracy was
achieved by very high vibration frequences of the quartz crystal.

The battery   problem was solved in the 1950s. Transistor
switching was devised in c.1953–4 and applied to watches in
c.1959. Integrated circuits became available in 1967–8. These
devices would eventually achieve all the complications of a
mechanical watch at a fraction of the cost.

Integrated circuits are microscopic in size (a 1973 integrated
circuit could contain 1,238 transistors on a 3.8mm ($\frac{1}{8}$in) square
chip of silicon) and it is not sensible to attempt to describe their
mode of operation; suffice it to say that the circuits count the
crystal vibrations and divide their frequency to levels suitable to
drive the digital or analogue display.

A digital display is achieved by exciting the correct 'cells' in
the 'double box' of seven cells used to portray the digits 0 to 9.
Conventional analogue display is achieved by use of a stepper
motor. Stepper motors have a rotor with six peripheral, alternate,
north and south magnetic poles. When the motor stator, which is
an electromagnet, is excited the rotor turns through one-sixth of a
revolution. When coupled with a 10 to 1 electronic 'gearbox', the
seconds hand will move in one-second steps and conventional
motion work arranges the rotation of the minute and hour hands.

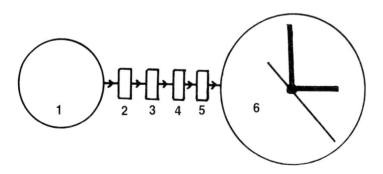

Fig 20 Electronic analogue display watch. 1 Battery 2 Quartz crystal oscillator
(32,768Hz) 3 Integrated circuit reducing frequency to 1Hz (1 tick per second)
4 Stepper motor   5 Hour and minute hand train   6 Display.

Electromechanical watches, which only needed the miniature battery to become feasible, operate by energising an electromagnet with a switch. The excited magnet sustains the balance vibration. Switching of the magnet was initially mechanical, but later transistor switching was used. The early watches of each type are described below.

Electromechanical watches with mechanical switching

1957 First battery-driven watch for sale, Hamilton 500 (US). Hairspring-controlled balance with a coil of wire on the rim. A mechanical switch operated by a jewelled pin on the balance roller energises the coil, and fixed magnets interact with the coil to impulse the balance. Hands driven by ratchet wheel.

1958 Lip watch for sale (France).
Hairspring-controlled balance with a mu-metal rim. Two electromagnets adjacent to the balance rim are energised from the battery by a mechanical switch on the balance roller. Impulse in one direction only, hands moved by a ratchet wheel system.

1960 Ebauches SA calibre L 4750 on sale (Swiss).
Similar to the Lip above, but the balance impulse is in both directions. Hands moved by a lever to increment the train. Sold in the UK using the brand names Avia and Mira.

Electromechanical watches with transistor switching

1960 Bulova Accutron watch for sale (US).
'Tuning fork' vibrator with a magnet at the free end instead of a balance wheel. The magnet interacts with a stationary wire coil in two parts: one part switches on an amplifier, and the other receives the amplified output to impulse the tuning-fork vibration. The hands are driven by an indexing arm (vibrating with the tuning fork) to rotate a ratchet wheel.

1967 Ebauches SA calibre 9150 (Dynotron) for sale (Swiss).
Hairspring and balance-wheel movement with electronic switching. The balance consists of two discs facing each other. Each disc carries two magnets, so that there are two magnetic fields between the discs. Two fixed coils in the gap between the discs switch on the amplifier to impulse the balance wheel. The watch train is driven by a jewelled 'lever'.

Electronic watches with a quartz oscillator

1967 Seiko hybrid analogue watch without a full integrated circuit (Japan).

1968 Ebauches SA Beta 21 prototype analogue watch (Swiss). The Beta 21 has a full integregated circuit with 8,192Hz crystal divided to 256Hz to drive a vibrating reed, indexing a ratchet wheel for an analogue display. On sale from 1970.

ESA Beta 21

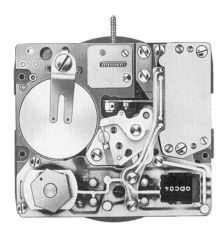

1970 Bulova Accuquartz analogue watch with integrated circuit (US).
Integrated circuit, crystal frequency 8,192Hz divided to drive a tuning fork and ratchet system as in the Accutron. Later models used a crystal frequency of 32,768Hz. The price in the UK was £1,250 in 1970.

1970 Hamilton Pulsar with light-emitting diode (LED) display (US).

1972 First liquid crystal display (LCD) watches from Longines and Société des Garde Temps (Swiss) and Seiko (Japan).
The Longines LCD watch used technology that came from Texas Instruments (US), but the early systems had a short life. This problem was solved in 1973 by Gray of Hull University (UK) by the use of a polarised light system. (This is simply explained by taking two polarised sunglass lenses and putting one behind the other. Look through the lenses and observe the effect as one lens is rotated. At one point in the rotation no light will pass.)

In the watch display, only the electrically excited dark cells can be seen. The only problem with LCD is that light is required to see the time and so most LCD watches have a push button to illuminate the display for night use.

Seiko Astron 35SQ

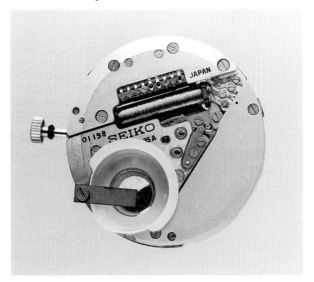

c.1994 Seiko Kinetic AGS (automatic generating system) (Japan). This quartz watch with analogue display has no battery. Instead, it has a swinging weight system similar to that in an automatic mechanical watch. In the AGS system, the motion of the weight operates a small electricity generator. The energy produced is stored in a capacitor and is used to drive a stepper motor for the display.

It is not envisaged that many readers will be interested in repairing electrical or electronic watches. For those who wish to enter this field, some useful books are included in the Bibliography.

CHAPTER 3

WATCHMAKING

Until the beginning of the nineteenth century, watch movements were made with simple tools and were relatively expensive objects. The developments in the eighteenth century discussed in Chapter 1 meant that the watch could become an accurate instrument and new ideas within the industry were aimed at producing watches using machinery. This would reduce costs and therefore enable a cheaper watch to be sold to a much larger number of customers.

EUROPE

The earliest successful attempt at using machinery was by Frédéric Japy of Beaucourt in France, who (in French patent 63 dated 17 Mars 1799) described ten machines for making watch ebauches (rough movements):

1 Circular saw for cutting brass sheet into strips.
2 Lathe for plates, fusees, barrels, collets, slides and racks.
3 Machine to cut wheel teeth.
4 Machine for pillar making.
5 Press tool to make balances.
6 Press tool for train wheels.
7 Vertical drilling tool.
8 Tool to rivet pillars.
9 Screwhead slitting tool.
10 Draw bench for the potence groove.

These machines enabled Japy to increase production to 40,000 per annum with 50 workers – this was a twelvefold increase in output per man.

At about the same time, the Swiss established an ebauche factory at Fontainemelon. An example of a Fontainemelon-style ebauche is shown below. This ebauche was probably used in Fleurier, where Chinese duplex watch ebauches were finished.

The next attempt to improve watch manufacture was that of Pierre-Frédéric Ingold, a Swiss worker who had spent some time with Breguet in Paris. Ingold had ideas for machines to make interchangeable watch parts, and in 1835 he attempted to form a French watchmaking company; however, opposition from the trade proved too strong. Ingold went to England to set up the British Watch and Clock Company in 1842 but, as in France, trade opposition persuaded Parliament that the industry would suffer if Ingold were to set up in business.

Left A French Japy Beaucourt watch sold in the late nineteenth century. The case is base metal and the movement is clearly a cheap design for machine manufacture. *Right* A Chinese duplex watch which beats seconds with a centre seconds hand. The Swiss ebauche is from the Fontainemelon factory and the centre for finishing this type of watch was Fleurier.

Ingold had issued three prospectuses in 1842 and had three UK patents: 9511 and 9752 (holding patents) and 9993 of 21 December 1843, which gave details of his machines. The story of Ingold's venture is told by R.F. and R.W. Carrington in *Antiquarian Horology* (see Bibliography). It is clear that Ingold made some watches and the author has seen three examples.

A more successful approach to machine watchmaking was that

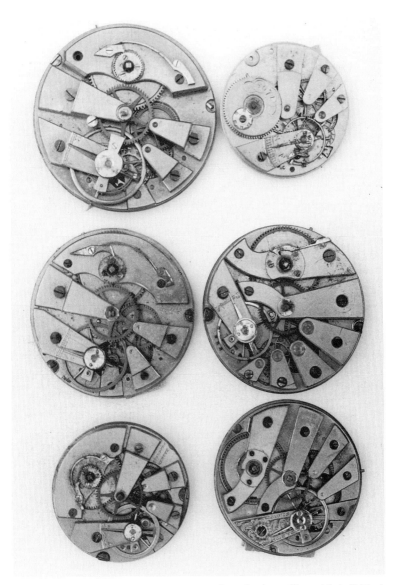

The development of Swiss 'bar' movements from Lepine calibre with individual cocks for each wheel. The top pair are collectable, as the left-hand movement has a temperature compensation curb and the right-hand movement is English, made in c.1840.

of Georges-Auguste Leschot, who was employed in 1839 by Vacheron and Constantin, Geneva watchmakers, to standardise their movement designs – so that standard-sized movements could be housed in standard-sized cases. This also allowed a degree of interchangeability in the parts. Leschot devised a pantograph machine which, from a master template six times full size, could perform a number of operations. The machine was used to drill holes in plates and bridges, to finish rough-profiled parts made in a press, to mill recesses and to perform engraving. His less well-known machines included hand presses, lathes for turning plates, and bridges with fixed tools on sliding blocks which had preset adjustable set screws, presses for wheels, barrel recessing machines etc. Leschot's achievements were to make Swiss 'barred movements' a very successful design for years to come. Many completed watches with such movements have attractive, multicoloured dials.

AMERICA

New developments in watchmaking took place in America. As early as 1837 there had been an attempt by the Pitkin brothers (Henry and James) to produce watches with interchangeable parts, but only about five hundred were made.

The real start in America was made when Aaron Dennison, who had been apprenticed to a clockmaker and later worked in Boston learning about watchmaking, met Edward Howard, who was a partner of David Davis in a factory making clocks, scales and weights in Roxbury. A new firm – Dennison, Howard and Davis – was set up in 1850 to make watches by machinery. Dennison brought Nelson Stratton with him from their watch and clock repair business. Significantly, Stratton had been an apprentice with the Pitkin brothers. The new firm had problems with dial-making and gilding, and both Dennison and Stratton went to England to seek advice. When Dennison came back to Roxbury, he set about making an eight-day watch, but it was not successful and a conventional 30-hour design was made. At first the watches used English-style escape wheels with pointed teeth, but the more robust Swiss-style escape wheel was soon adopted.

The company name changed frequently, to become the American Horologe Company, the Warren Manufacturing Company (which made 100 30-hour watches engraved 'Warren') and in 1853 the Boston Watch Company. The next 900 watches were engraved 'Samuel Curtis'. In 1854 a new steam-powered factory, together with houses for the workforce, was built at Waltham. The 100 workers produced 30–40 watches per week.

One of the crucial decisions in this early period was to employ an engineer, Charles Moseley, who could understand the problems of machinery, measurement and accuracy. However, Moseley left in 1852, only to be re-employed in 1854 when his new company was acquired by the Waltham-based firm. This somewhat shaky start, with associated development costs, in turn caught up with the Waltham-based company, which became bankrupt in 1857 and was sold by auction to astute investors who realised that the future was not so bleak. The gold watch-case makers Tracy and Baker joined with Royal Robbins and D. F. Appleton to acquire the firm, which was renamed Appleton, Tracy and Company. There were further name changes to the American Watch Company in 1859, the American Waltham Watch Company in 1885, and in 1906 the Waltham Watch Company. Robbins was in charge from 1857 to 1892. Dennison was discharged in 1862 after an argument about a new cheap model to be known as the William Ellery. Ironically this model so keenly advocated by Dennison was in fact produced, and by 1865 accounted for 45 per cent of sales.

Howard had no place in the new company and was in debt, but concurrently with the developments outlined above he set up in the old Roxbury plant trading as Howard and Rice. Later, after clearing his debts, the firm became E. Howard and Company. With success available for all to see, other companies were formed, but only a few survived. (A list is provided on page 88.)

The most important outcome for American makers was the realisation that modern mass-production methods were an engineering enterprise and that interchangeable parts could be used for both watchmaking and watch repairing. It is also interesting to note that American watches made by machinery in standard sizes could be cased in any style or make of case. This could be chosen by the purchaser, so that an attractive case might easily contain a more mundane watch and vice versa, depending on the purchaser's choice.

Later, a new 'cheap' watch industry emerged. It was led by an interesting design made by Daniel Buck, which was intended for a section of the market unable to afford better watches. This is known as the Waterbury watch and cost $3; it had a pressed-out duplex escapement (US Patents 20,339 and 20,340 of 1878). Early Waterbury watches had a 3m (10ft) long mainspring in the back of the case, which drove the whole movement around a fixed centre wheel once per hour. Series A to E Waterbury watches have this design; the later series watches F to W have a more conventional movement but they still retain the pressed-out duplex escapement.

The American industry had also observed the effect of Roskopf's pin-lever escapement design and the associated developments in Switzerland. As a result, American firms now entered this market, the first being the New Haven Clock Company in 1880, closely followed by the Waterbury Clock Company associated with Robert and Charles Ingersoll. Basically, the early Ingersoll watch was a small pin-lever clock movement which was fitted into a large watch case, the folding clock winding bow being inside the back of the case. From 1896 this was sold for $1. This led to a 'Dollar watch' industry. The Ingersoll company sold 96 million pin-lever watches, and by 1930 the total US output of approximately 272 million watches included 186 million Dollar watches (no longer necessarily costing one dollar). Other well-known Dollar watch companies were the Ansonia Clock Company, the Western Clock Company (Westclox) and the Bannatyne Watch Company (E. Ingraham). There were also other short-lived firms.

Ingersoll in America went bankrupt in 1922 and the Waterbury Clock Company purchased the remains. Ingersoll in England became a separate company. In 1942 the Waterbury Clock Company was acquired by a Norwegian refugee, Joakim Lemkuhl, and during World War II the company made fuses. After the war they were renamed the United States Time Corporation, and using armalloy bearings and pallets for long life, and with a superior design of pin-lever escapement, they produced watches. These were called Timex.

Timex became a multinational company. Their first overseas subsidiary was in Scotland, where they initially branded their products Kelton, and by 1972 they were making 30 million watches a year in various parts of the world. In due course they produced electronic watches, which can still be purchased today.

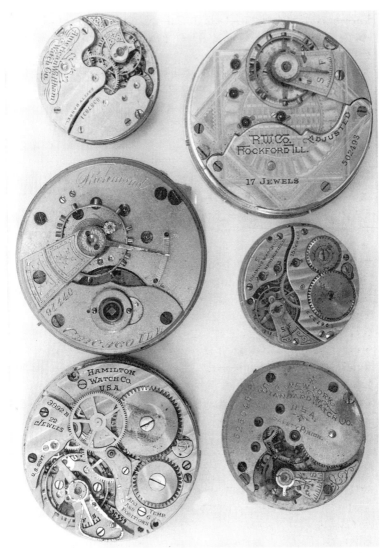

A selection of machine-made American movements. The makers are: *top* Waltham and Rockford; *centre* Hampden and Illinois Watch Company; *bottom* Hamilton Watch company (military watch) and New York Standard Watch Company.

Westclox also survived as a multinational company, again with a factory in Scotland.

Finally, what of Aaron Dennison, dismissed from Waltham in 1862? Dennison had more watch-company ventures in America, including the Tremont Company from 1864–6 and the Melrose

Watch Company from 1866–8. In 1871 he became manager of the Anglo-American Watch Company in Birmingham, England, but this firm was wound up in 1874. The remains were purchased by William Bragge for £5,500 and renamed the English Watch Company, but Dennison had no place in it. Dennison then decided to make watch cases and founded the well-known Dennison Watch Case Company (again in Birmingham); this was part of the English horological scene until 1967.

Major American watchmaking companies listed in approximate order of formation

Waltham	**Waterbury**	**US Waltham**
Howard	**Peoria**	**New York Standard**
US Marion	**New Haven Clock**	**Hamilton**
Newark	**Columbus/**	**Ingersoll**
Elgin	**SouthBend**	**Westclox**
Hampden	**Cheshire**	**Ansonia**
Illinois	**Trenton**	**Ingraham**
Lancaster	**Manhattan**	**Manistee**
Rockford	**Aurora**	
Auburndale	**Seth Thomas**	

Details of these companies may be found in the National Association of Watch and Clock Collectors Inc *Bulletin Supplement* (see Bibliography).

Effects in Europe

The effect of American watchmaking methods was to galvanise the Swiss industry. Specifically it increased the use of machinery, although the process was not so comprehensive as in America which had no traditional industry. One of the most significant Swiss inventions for watchmaking in the last quarter of the nineteenth century was the sliding-head automatic lathe, which eventually became the mainstay of modern mass production of small parts for any purpose.

The main 'external' change in a Swiss movement design was the increasing use of various shaped bridges and plates instead of the Leschot bars. The Swiss also became most adept at making watches to suit the customer. Many apparently American or

English watches were Swiss made, and if they are are not marked as such, removal of the dial would confirm their origin. Swiss parts eventually became interchangeable but this was a slow process, probably due to the traditional employment of outworkers. This is indicated by a list of 713 Swiss 'makers' in *Kelly's Directory of the Clock, Watch and Jewellery Trades* for 1901. It was not until the 1960s that the advent of electronic watches forced a change to manufacture in suitable factories.

UK

English watchmaking was also a labour-intensive industry, but in a different manner to that of the Swiss. The English industry was frequently based in small workshops largely in London, Prescot and Coventry. It is significant that the total English output was constant, whereas Swiss and American output was rising:

	English	Swiss
1800	200,000	200,000
1850	200,000	2,200,000

In fact, the English industry was in terminal decline.

If a late nineteenth-century 'handmade' English watch movement is examined, there will often be a name on the visible plate. For some watches this will be the name of the maker, but usually it is the name of the retailer. If the dial is removed there will often be numbers. One number may duplicate that on the visible plate, but there are usually others such as 14 03 or $14\frac{0}{3}$ which represent the watch size (see Appendix on page 177). There may also be initials, such as the relatively common JW or MM(see next page). These are the initials of the rough movement (ebauche) maker, both of whom were working in the late nineteenth century.

In England, the watchmaker had traditionally employed different people, each of whom would perform his own particular task; thus the movement may have travelled to various premises. The maker of the watch co-ordinated the work and provided the final finishing process. *Rees's Cyclopaedia* of 1819–20 lists 13 workers to make a rough movement, and 21 to finish it; many of these would have worked at their own premises. This method continued until the latter years of the nineteenth century, although there was an increase in people who worked together in one small workshop under the control of the 'watchmaker'.

The author is fortunate to have been allowed to examine the records of a rough-movement maker and a small watchmaking firm, both of the same late nineteenth-century period. An analysis of the information is not entirely satisfactory, since not all of the 100 gross of frames made in a year by the movement maker were sold to the one watchmaker, whose records show that he made and sold only 200 watches each year. This watchmaker employed six people directly and purchased balances, caps, escapement parts, hairsprings, glasses, hands, dials, fusee chains, jewel holes, jewel screws, regulators and so on. He also had the watches gilded, engraved and cased. This represents a total of about twenty people involved in making a watch.

The initials on the frames mentioned earlier are those of John Wycherley (JW) and James Berry, with (MM) supposedly meaning 'machine made'. James Berry also traded as Isaac Hunt. Many of the English watches available to the collector will show these initials.

John Wycherley was an important pioneer in the process of introducing interchangeable parts in England. Nevertheless, however interchangeable his frames were, the methods used by the 'maker' would tend to make the parts unique to the watch. On 26 March 1867, Wycherley was granted UK patent 880 for:

> Improvements in the mode of and Machinery for Manufacturing Watches, Part of which Improvements is applicable to Lathes generally where great Accuracy is required . . .
>
> The chief object of this Invention is to effect an economy in the manufacture of the parts constituting the framework of watches, and to form them with such accuracy as to render the parts interchangeable instead of (as is now the practice) making and fitting the parts for each watch separately. There are six pieces which go to form the framework of a watch, and it is these pieces which I design to conform to standard gauges by the means to be presently explained. The pieces are, (1) the pillar plate; (2) the upper or top plate; (3) the third wheel or pillar plate bar; (4) the pottance; (5) the cock; (6) the name bar . . .

The remainder of the patent describes the machines and processes used.

There was a further stage in Wycherley's innovations in that he would supply the frames undrilled for the wheel arbors but marked with the exact position for the hole to be drilled for the wheels to engage correctly. This innovation was not universally acceptable to the customer, but if the user preferred to plant the wheels himself he could still do so. Wycherley was able to supply his rough movements as separate packets of frames, wheels, bridges etc, instead of an assembled entity, *and* it would be possible to supply spare parts which would fit when pivoted by the repairer.

The Coventry firm of J. Rotherham tested some of Wycherley's movements, and wrote a letter to be read out at a meeting of the British Horological Institute on 12 September 1867:

Coventry, September 5th, 1867

Dear Sir,

We received from you seven frames and one set of materials. One of the frames had holes in it in the old style, and we pivotted the materials to that frame in the usual way running the depths with an ordinary depthing tool. The other six frames had no holes in them, but had small marks made by you in the places where you considered the holes should be. We gave one of our men the one set of materials and the six frames, and told him to drill the holes where you had marked them, and then to open the holes and try the set of materials, in all six. He did so and found the heights and depths quite correct. We then put the top plates on different pillar plates, and still found the result the same... We have since had a rim-cap made to one of the first gross that you have sent us of movements on this plan, and we find that it will fit every frame in the gross ...

Yours &c,
ROTHERHAM & SONS

English watches were usually made in one of three centres: London, Prescot or Coventry. Thirteen years after Rotherham's comments, a letter from C. A. Read appeared in the *Horological Journal*, 22 July 1880, describing the method of watchmaking in Coventry using Wycherley's movements.

The letter explains that in Coventry there were three stages in making a watch, and describes the work involved in each stage. These were sometimes known as first half, second half, and finishing. Examining is also mentioned, but it was clear in the records of the small maker whose books were studied that finishing and examining were arranged flexibly to suit the workload. The workers in the firm were described as first half, second half and finisher or examiner.

The firm also had many movements which were not completed when it closed. These had been stored in 'tin boxes' approximately 5cm (2in) high and 5cm (2in) in diameter. It was possible to follow the progress of watchmaking, since each was numbered and the date of each stage was recorded. This firm was probably not unique in such actions, for there would be little point in finishing watches until an order was received. By American standards, however, this was clearly an outdated industry.

Extracts from Read's letter are shown below:

...The minor parts are principally done by apprentices; the old system of putting a boy to polishing up when he first goes apprentice, and keeping him at that for a year or more, and in many cases two or three, has been found to be a mistake, and the polishing up is principally done by girls, and in some cases the finishers learn their wives to polish up...

The first thing a boy does now, after having the usual day at filing pins, and a few days at turning down pieces of steel, is to learn to do fuzees, then barrels, and after, centre wheels. He is then kept some time doing these three things. Here we get the first division. The next is the third and fourth wheels. That is always made a separate part to pivot and polish, when they are not to be gilt, Wycherley's movements are generally used, and they are sent out – so many movements [frames] so many third and fourth wheels, so many fuzees, barrels and centre wheels – all in separate boxes. They are all the same for a certain size, so that it does not matter about mixing the wheels. Two or three dozen third or fourth wheels are given out at a time to a man, or apprentice, as the case may be. By doing this

Top A pair of English rough movements as delivered from the maker (T + S). The protruding wheel arbors are in oversize holes in the plates, and the arbors are not pivoted. Similarly, the fusee is uncut. *Bottom* A verge watch in a movement tin, which started life as an ebauche similar to those above. Work has been done so that the wheel arbors have pivots, the fusee is cut, and various other parts have been processed or bought out.

one thing for a time, a person gets very quick at it, and can also do it much better. I know a man who has done nothing else for at least ten years, only to pivot third and fourth wheels...and earns good wages... The wheels are all pivoted to a gauge, and the movement is not used in the pivoting at all... The next part, is what is called in the trade running, and this is the first time the movement is brought out and put into the usual tin box. One or two dozen are given out at a time... The third and fourth wheels are brought in by the pivoter... Fuzee, barrel and centre wheel are also put in each box, all ready pivoted, with detent and stop work, chain, spring, barrel, arbor, ratchet, click and screw. The runner's work is to drill the holes, the depths being ready marked by the movement maker, and run in the wheels... He also pivots in the detent, does the stop work, hooks in the mainspring and tries the adjustment. The next thing is the balance staff... The difficulty of fitting the roller is got over by always using one maker's escapement, and getting him to make the holes in the rollers all one size.

Next comes the pitching of the escapement. The staff pivoter, having taken the watch in with the staff in the frames, the escapement is put into the box. Having been already pivoted to gauge, same as third and fourth wheel, by another man called the escapement pivoter, the escapement pitcher has only to pitch the escapement, and see that it is all right. The watch then goes to the finisher whose work is now diminished to a very small portion. He first studs and springs it, then blue stones it up and gets the watch gilt. He has then only to put it together. The next and final part is the handing and examining. The examiner is supposed to examine the watch all through, hand it, time it, and pin it in the case.

It is clear that a small firm could not employ as many people as Read mentions, and unless he sent work out, his own workforce would need to be more flexible.

English watchmaking did eventually turn to machine manufacture and, as mentioned earlier in the Dennison saga, the first company was called the English Watch Company; this was set up in 1874 and purchased in 1876 by William Bragge. The new company acquired the remains of Dennison's final watchmaking adventure, which had been as manager of the Anglo-American Watch Company.

English watchmaking companies

The *English Watch Company* would at first have needed to buy out parts, since the Dennison company only had the ex-Dennison plate-making machinery. The new company was listed on the

One full-plate and one threequarter-plate under-dial view for the identification of English factory-made watch movements: *top* English Watch Company; *centre* Ehrhardt; *bottom* Rotherham. (See also the photograph 98.)

stock exchange in 1882, but William Bragge died in 1885 – the year that the factory was enlarged. At this stage they were probably making most of their parts, including English pointed-tooth escape wheels rather than the Swiss style. The EWCo also purchased the rights of Mr Douglas of Stourbridge to make chronographs. It is thought that this would include those in patent 4164 of 27 September 1881 which arranged for the fitting of a centre seconds hand and a minute counter to a normal watch.

The company went into voluntary liquidation in 1895, having made about 200,000 watches, most of which were key wind. The company watches may be recognised by their trademark, a monogram of EWC or the words 'Haseler's Patent' on the pillar plate of some models. Occasionally, watches are explicitly marked 'English Watch Company' on the top plate. One peculiarity of EWCo products is that size is usually represented by a letter on the top plate near the serial number. The code is roughly S = 12 size, R = 14, W = 16, N = 18, V = 20 and E = 22 (probably due to the use of ex-American machinery).

William Ehrhardt of Birmingham was a German who came to England when he was 20 years old, but he was later 'naturalised'. The Ehrhardt firm made about 800,000 watches between 1876 and c.1923. William had died in 1897, but his sons continued the business. In 1874 Ehrhardt had a factory built in Birmingham known as Time Works, which is probably when he set about machine manufacture in earnest. Ehrhardt watches may be recognised by a winged arrow trademark on the pillar plate, and there are also a few explicitly marked 'William Ehrhardt' on the top plate. Sizes range from 8 to 20, and both full-plate and three quarter-plate watches were made.

Rotherham & Sons of Coventry had been in business from the mid-eighteenth century, but their interest in machine manufacture started in 1856 when John Rotherham (1838–1905) visited America. In 1880 Rotherham purchased machinery from America, and by 1890 it was reported that the firm produced 100 watches per day. It is the author's view that the factory may have been capable of this, but that the achieved output would have been dictated by sales.

Serial numbers suggest that Rotherham made about 430,000 watches between 1880 and 1920. Full-plate and threequarter-plate models were made; sizes range from 8 to 20. Rotherham watches usually have a trademark showing a serpent coiled around a vertical rod, the whole being surrounded by a star-shaped frame. It is thought that Rotherham may have supplied parts to other firms such as Benson, and Smith & Sons Ltd.

The *Lancashire Watch Company* was set up in Prescot in 1888, soon after T. P. Hewitt had acquired Wycherley's ebauche-making business. Hewitt continued to make ebauches at the new watch company for many more years. However, he was intent on acquiring as many as possible of the small Prescot ebauche-making businesses, and if he could, he paid them in LWCo shares. When the factory being built for the company was partially complete, he re-sited movement-making into the premises, and between 1890 and 1893 the company sold a total of around 300,000 ebauches.

The first watches started to appear in 1893 but the venture was too late, with dated designs. The company made about 900,000 watches with Swiss-style escape wheels, with just a few with English style, but the business was always fighting a losing battle against imports and the products of Williamson (see page 99). Towards the end, the LWCo accepted the need for cheaper, modern-looking watches and produced the Vigil, a threequarter-plate lever watch, and a pin-lever model called John Bull, which sold retail for five shillings. The John Bull watch does not have a recognisable numbering system, but the company records suggest that about 5,000 were sold.

In 1908 receivers were appointed, and in 1911 the remnants of the company were sold by auction.

The *J. W. Benson* story starts in 1749, but little is known of it until 1840. Benson were both makers and retailers, and in 1892 they built a steam-powered factory at Belle Sauvage Yard in Ludgate Hill, London. It has been suggested that they did not make watches – but for what other purpose would they need such a factory?

Correspondence with Benson some years ago resulted in a positive statement that they did make watches, but they also agreed that other makers' watches were also supplied. Since they were retailers they would need a varied selection of watches to sell, and this would include not only other English products but also Swiss and possibly American watches. Evidence of manufacture is contained in a report in the *Horological Journal* of a visit to the factory in 1935 (see Bibliography). Their own products were the

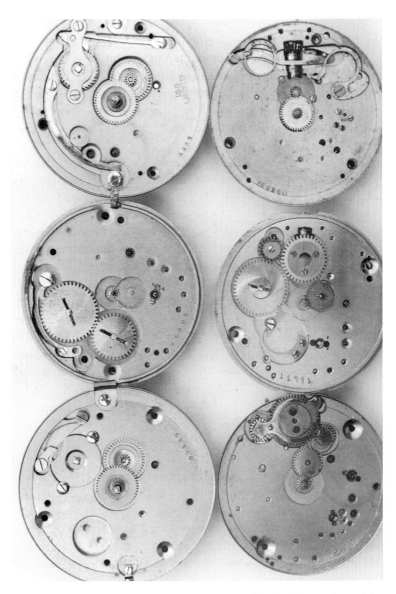

This photograph is similar to that on page 95 for identification: *top* Lancashire Watch Company; *centre* Benson (both threequarter plate) – *left* key wind, *right* keyless; *bottom* H. Williamson products.

Ludgate, Field, Bank and the 'Best London Make' models. It is also known that Benson bought watches from Guye and Nicole Nielson in England, and various Swiss watches.

The Benson factory and records were destroyed by bombing in the 1941 Blitz, so their total output is unknown.

H. Williamson Ltd was a serious attempt to establish a 'machine-made' watch factory in Coventry. In 1895 Henry Williamson acquired the small watchmaking business of Charles Huton Errington, and in 1898 Williamson became a Limited Company. At roughly the same time Williamson either founded or purchased a business at Buren in Switzerland to make parts for his watches in England, using frames made in Coventry. However, the Merchandise Marks Act of 1898, in force in 1899, caused him problems. A group of English watchmaking firms, led by Hewitt of the Lancashire Watch Company, claimed that Williamson was using foreign parts in watches described as 'bolt and cap English Levers, etc'. Williamson lost the case but not the battle, for he enlarged his Coventry factory to make parts and was able to watch others fail as he became successful.

Williamson made full- and threequarter-plate, key-wind and key-less models, culminating in a very successful three quarter-plate model which supplied the government with up to 1,000 watches per week during World War I. After the war there was more success, but between 1924 and 1931 the business declined and by 1930 the firm was bankrupt, having made over 600,000 watches.

More detail on English watchmaking factories may be found in an article by myself in *Antiquarian Horology* (see Bibliography).

Other watchmaking initiatives

It should be clear that the total annual output of these factories was not enough to satisfy the UK demand, and cheaper Swiss and American products continued to be imported in large numbers. English watch manufacture had virtually ceased by 1930, which was not in the national interest, and before World War II the government encouraged S. Smith and Sons to start reviving a watchmaking capability. Smith's acquired both machinery and material from Switzerland and managed to make clocks and watches at Cheltenham for aircraft use, together with pocket and stop watches for the military. After the war, the government decided to expand the watchmaking industry, to avoid using valuable foreign currency for imports to satisfy a pent-up demand after six years of war.

S. Smith and Son, Ingersoll (UK) and, for a short time, Vickers Armstrong were one group, and Newmark of Croydon were another. Smith's made jewelled lever watches in Cheltenham and the Smith Ingersoll Vickers Armstrong group built a new factory at Ystradgynlais in Wales to produce pocket and wrist watches with pin-lever escapements. The pocket watch was basically the old Ingersoll Crown model. These watches were sold under three labels: Smith, Ingersoll and Services. Later, Smith's imported movements from Japan, including electronic models, even though they had in fact developed their own electronic design called Quasar which was not put on sale.

The Ingersoll venture in England commenced when Simonds London Stores ordered a million watches to be delivered in 1900–3. Simonds ceased trading in 1904, and Ingersoll opened their own branch in 1905, which was followed by an assembly plant in London in 1911. In 1915 this English branch became Ingersoll Watch Company Ltd, which by 1922 was assembling 3,000 watches per day from American parts. In 1922 the US company were bankrupt, and in 1930 the new owners (Waterbury Clock Company) sold the assets of the UK company to the London directors. A factory was opened in Clerkenwell, where assembly of American parts continued. The American Ingersoll wrist watch was too large for the European market, and the UK company decided to import wrist watches or movements for casing from Switzerland and Germany. Many different models were sold, at prices ranging from 7s 6d to 30s; brand names include Leader, Legion, Rex, Sports and Elite.

None of this represents manufacture but, as indicated above, watches were made in Wales after World War II. The Ingersoll brand name continued to be used after their withdrawal from the Welsh venture in 1969. Smith's continued to produce pin-lever pocket watches in Wales; they also made three pin-lever wrist-watch models, and a total of about 30 million watches were produced before closure in 1980.

The Services brand mentioned above was sold by the Services Watch Company Ltd of Leicester, who were not watchmakers. They were in business from the 1930s to the 1970s. Before World War II they imported wholesale, and supplied retail shops with Swiss and German pin-lever pocket watches and also wrist watches.

Services and Ingersoll (UK) watches before World War II. *Top left and centre* Thiel; *top right* Junghans; *bottom* EGM 8½ ligne. Pieces of string from tie-on identification labels may be seen in this and some of the other photographs in the book. Tie-on labels are preferable to sticky labels, which may damage some case materials.

The 'patriotic' brand names included Army, Airman, Despatch Rider, Marine and others, but the company's supplies would have ceased during the war years, and they were no doubt employed for essential work in that period. After the war they resumed their watch business with Welsh products, but as soon as imports were available they resumed their normal business but with better-quality products.

Newmark watches were not part of the Welsh venture, but they too had government support to open a factory in Croydon at which pin-lever watches were made. Newmark had been wholesale importers of Swiss and German watches since 1875, but had not resumed German imports after World War I. Approximately seven million Newmark watches were made up to 1960, when they resumed their importing business. Their main suppliers are Avia and Swatch.

In 1947 Timex opened a watch factory in Dundee, and Westclox opened a clock factory in Strathleven in 1948. The first Scottish Timex products were labelled Kelton and contained a modified Ingersoll movement. The first Timex design was called the Kelton 21 in Scotland and was made until 1959, after which the label Timex was used. This is not the place for a history of Timex world-wide watch-production methods, but it marked a complete change in watchmaking ideas. Manufacture was established wherever Timex felt it made economic sense. Similarly, the Timex marketing strategy of allowing their watches to be sold by any suitable outlet willing to sell was not a popular idea in the American jewellery trade. At first retailers refused to sell Timex watches, although customer demand eventually forced them to change their views.

The Westclox clock factory in Strathleven was expanded on an adjacent site in 1959, and the new factory was used to produce a pin-lever wrist watch based on a mid-1930s design. The rectangular movements were stamped 'Made in UK', but the complete watch was labelled 'Made in Scotland'. In 1969 Westclox became part of Talley Industries, who continued to market the mechanical watch until the advent of cheap electronic watches reduced the demand for their products.

WATCHMAKING IN OTHER COUNTRIES

France

Watches have been made in France since c.1550; there is a watch of this date in Paris, by Jacques de la Garde. Paris was also the location of Abraham-Louis Breguet, probably the world's greatest watchmaker, for much of his life. Then there is the contribution of Japy, who changed the methods of ebauche manufacture discussed at the beginning of this chapter. The 1901 *Kelly's Directory of the Clock, Watch and Jewellery Trades* shows that there were 112 watchmakers in the Besançon–Morteau area of south-east France. The Japy company is listed, and evidence in advertisements of the 1920s shows that the Japy firm was still making watches. In 1939 French watch production was about two million per year; after World War II the French industry made about six million watches per annum, which were sold in both Europe and America.

Germany

The German watch industry was relatively sparse until the end of the nineteenth century. About 5,000 quality watches per year were made at Glashütte near Dresden, the best-known firm being A. Lange and Sohne. Germany also became a pioneering manufacturer of pin-lever watches between c.1885 and 1939, the main firms being Thiel, Haller, Kienzle, Junghans and Müller-Schlenker. After World War II, the industry was revived in both East and West Germany, so that by 1976 East German production was about six million per annum and West German nine million. The only East German watch I have seen was branded 'Ruhla, Made in GDR'.

One very important contribution made by Germany to watch history is the production of the Flume *Werksucher*. This book contains illustrations of the under-dial features of all watches for which spare parts could be obtained in 1947, and is invaluable to collectors for identifying movements. (There are other, similar books available – see Bibliography.)

USSR

Russia was not a traditional watchmaking country but, presumably for strategic reasons, they purchased the bankrupt Ansonia Clock Company (US) and also the Dueber-Hampton Watch Company (US); this firm made jewelled lever watches (Dueber-Hampton was the last name used by the Hampton Watch Company in the period 1923–30). In this latter case, they not only bought the company equipment but also hired 21 workers to set up the new factory and train the Russian workforce. It is stated that in 1937 a three-shift routine, which employed 3,000 workers, made 20,000 watches per week.

After World War II the Russians established a modern watch industry. All watches seen in Britain are labelled Sekonda. Mechanical watches made in Russia have 'USSR' under the 6 number on the dial, but quartz watches using imported movements do not have the 'USSR' mark.

Japan

Traditional Japan did not use the same timekeeping system as Europe, America and other countries. However, in about 1860 conventional watches using imported Swiss movements were made. A factory was set up in 1894 in Osaka to make pocket watches using machinery obtained from the defunct Otay watch company of San Diego (1889–91), and about 19,000 watches were made.

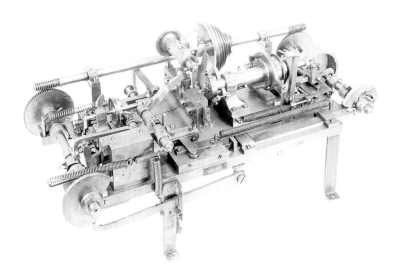

The invention of automatic lathes in Switzerland to mass produce small round parts such as screws, arbors and so on took place in the late nineteenth century. This Swiss lathe was made in c.1885. The tools were in a fixed, longitudinal location, and the metal being processed was moved through the plane of the tools. As each part was completed it was cut off and the operations were repeated. The necessary longitudinal movements of the work and lateral movements of the tools were achieved by cams fixed on the camshaft on the right-hand side of the picture.

In 1881 Kinttaro Hattori founded a company in Tokyo, and in 1892 he started making clocks, and later pocket watches, at the Seikosha (Tokyo) plant using the brand name Time Keeper. The Hattori movement looked like a late 'bar' movement of Swiss design. In 1913 Hattori made a 12-ligne, 7-jewel wrist watch branded Laurel. After the 1923 earthquake, in which the Seikosha plant was destroyed, Hattori rebuilt the factory and in 1924 made a wrist watch for men with the brand name Seiko. By 1936 production was about 3.5 million watches per annum, a figure not reached after World War II until 1956. By 1987 production had risen to 100 million watches per annum. Seiko also make cheaper Lorus, Lassale and Puresar watches.

The Citizen Watch Company (Shokosha Watch Research Laboratory) exported watches from 1936 and by 1984 was producing 55 million watches per year.

Hong Kong and China

Hong Kong imports movements for use in either imported cases, or cases that are made in Hong Kong; the output in 1985 was 325 million watches. Chinese watches using the brand names Seagull and Double Rhomb appeared in Britain in 1977–8, and Swiss estimates suggest that China 'made' 35 million watches in 1985. It is not known how many of these were electronic. The recent return of Hong Kong to China will probably result in a larger production rate.

One interesting facet of Far East watchmaking is shown on an inexpensive analogue electronic watch purchased as a present for a child. The information given on the dial was the brand name, Constant, and 'Japan movt'; however, on the steel case back, the words 'Assembled in China' appear. So where did the case and strap come from, and where was the watch 'made'?

WATCHMAKING TODAY

It would seem that by 1998 the Swiss watchmaking industry had achieved a miracle of survival against the financial odds. A study of catalogues made specifically for this chapter suggests that mechanical watches have become a fashion accessory and status symbol. This is good news for Switzerland, quality watchmakers and future collectors. However, the timekeeping and available functions of a complicated mechanical watch may also be achieved in a cheap, digital, electronic watch, so that similar features can cost £15 or £15,000! No discussion of the manufacturing processes for such electronic watches has been attempted here, because these are entirely different to those described for mechanical models; the *principles* of such watches have been outlined in Chapter 2.

The advent of machine-made watches does help when 'mending' such watches, because parts can be taken from old movements. This means that many of the problems of repair are only a matter of locating spare parts. At this stage, many years after the watches were made, this involves finding an 'identical' watch or movement. While this is sometimes successful, there are problems because manufacturing firms change their products from time to time. This means that you would need to know to which 'batch' your watch belonged, just as for car spare parts you need to know a chassis or engine number. A simple horological example would be a small change in pillar height by the maker, which would cause major problems for a repairer 50 years later.

CHAPTER 4

COLLECTING

Collecting is a personal matter and in this chapter I will outline my own experiences: it will be seen that there are pitfalls!

In my childhood, small boys collected stamps, cigarette cards, bird's eggs and shrapnel. As an adult, I had no interest in collecting anything until on a wet day in 1968 I happened to walk into Lancaster museum and was confronted with a display of verge watch movements. On my way home I stopped in Tewkesbury and found a jeweller's shop which had a verge watch for sale. Since then I have been 'hooked'.

This experience will not be unique, and in my case has led to 30 years of horological *fun*, not always so much concerned with collecting but with research into the history of watches and watchmaking. As already mentioned in the Introduction, a few years later I found that I was more interested in movements than complete watches, and I disposed of all of mine, except for the Tewkesbury verge and the John Bull featured on the jacket of this book, the latter being the last one I purchased.

Although I now concentrate on movements, I photographed all the watches I owned before they were 'recycled'; today, the photographs are nostalgic souvenirs of times past as well as being a useful source of illustrations! I also found that much of the fun of collecting was concerned with finding a collectable watch, so that resale later was not heartbreaking.

My first watch was a good example of bad collecting, but fortunately with a satisfactory end. The watch was engraved 'Ja. Powell, watchmaker to the Prince Regent, 1137 Worcester'. Investigations revealed James Powell listed in Baillie's *Watch and Clockmakers of the World* (see Bibliography) as 'Worcester, 1804–8, watchmaker to the Prince of Wales, watch [in] Horstmann College'. The silver hunter case had the casemaker's mark 'ROJE', which indicated that it was made by Richard Owen and

James Edwards, and was hallmarked 1854. My watch movement was not in the original case! The movement quality and engraving suggested that it was originally in a gold case which at some time had been scrapped for the bullion value. This was a good lesson to learn, for subsequently it became clear that interesting movements are sometimes sold in the wrong cases, and that it is possible knowingly to buy a good maker's working movement in a married case which has protected it from damage.

Three years after my first purchase I had acquired more verge watches, but I slowly realised that there was a lack of variety in my small collection. I was again fortunate in buying a box of about thirty movements of various types, and I spent some time taking them to pieces to see what they were and how they functioned. After this I tried repairing the more interesting ones. There were no verge movements complete enough to repair, but there was a variety of lever and cylinder movements which I was able to restore to 'ticking' condition, using the methods discussed in Chapter 5. The most common faults were broken pivots, broken fusee chains, faulty clicks inside the fusee bottom, 'mis-shapen' balance springs and broken mainsprings. Later experience showed that these movements were not alone in having these faults, which in itself was a useful lesson. I never made any attempt to repair broken balance pivots, and on the few occasions that I have found this to be necessary I have used a professional repairer.

These experiences demonstrated to me that in order to avoid purchasing errors it was essential to have an organised assessment routine for a watch. One such routine is suggested below, but it is neither exhaustive nor foolproof and should be regarded only as a guide.

Watch assessment routine
1 Do not be rushed into a purchase.
2 Check whether you can change your mind within an agreed period.
3 *Examine* the external condition. Look at the material, condition and hallmark date on each part of the case. Is it numbered (not always on all parts)? Remember to check the movement number with the case number. Is there any evidence of a winding-key hole being filled? If gold is the material, what is the carat value? If you

can, try to assess the case weight – ie how much is the gold cost-ing you, and is this what you wish? Is the glass original, or if replaced, is it the right style for the date? Is the bow style correct for the date? Is the dial style correct for the date? What is the state of the case hinges?

4 *Examine* the movement condition. Open the front of the case and examine the dial carefully to check for any damage, *including hairline cracks* – dial repairs are difficult and cracks affect value. Does the dial have a name or number on it? The name may be that of the retailer or the maker. Check the dial number or name with those on the top plate of the movement (you will probably want to do this later in the examination). Do any of the numbers match the case number? Are the hands correct for the date of the watch? Are they replacements? Are they damaged?

5 *Examine* the hinge of the movement: is it a good fit? Remove the dust cap (if fitted). Is the movement full, half or threequar-ter plate? Remember to check the movement number and engraved name for agreement with earlier data. What sort of escapement is fitted? Is it the original? Some movements are gen-uinely 'modernised'. Are there any unusual features?

6 What is the state of the gilding? Are there scratches around the pins or screws? Are the screws blued or polished, and are they all the same? Are they damaged in any way? Look at the pivot holes in the top plate: are there any signs of wear or repair? Look at the jewelled holes: are any jewels cracked.

7 What sort of balance does the movement have – plain, steel or gold, or compensated? Are there any unusual features in the layout? Does the watch have any special features such as chronograph, day/date/moon phase, or repeating? If so, do they operate correctly?

At this stage you may come to a decision, but even if you feel that you will not buy the watch it may be worth continuing.

8 It is now time to see how well the watch appears to work, so put it all back together and *ask the vendor* to wind it, so that you can check that it works in various positions (dial up, dial down, verti-cally etc). Listen to the 'tick' in each position, and *ask the vendor* to re-open the case so that you can observe the balance motion: does it look right, or is it 'wobbly'? Do a five- or ten-minute check to see that the hands are functioning properly, observing the time elapsed on your own accurate (electronic?) watch. Wind the watch a little more yourself, test the handset mechanism through at least one hour and preferably 12: a tooth missing on a wheel or pinion of the motionwork is likely to cause problems.

9 Check anything else you wish and decide how much of an asset the watch will be to your collection, and how much it is worth *to you*.

10 Finally, if the watch is not satisfactory, *thank the vendor* and walk away – unless you feel that you will never find such a watch again, in which case you have a decision to make. On such occasions, whatever your assessment tells you, your judgement may be clouded and you will buy the watch anyway. This is simply a hazard of collecting, but see point 2 above.

One exception to the negotiated safety factor of 'changing your mind' is purchase at auction, because unless there is a specifically misleading statement in the catalogue, the watch is yours at the fall of the hammer. Remember that there will be auctioneer's charges and VAT to pay in addition to your successful bid.

When you have studied the above assessment routine, it might be worth buying a few cheap movements to examine, take to pieces, reassemble and so on. This should give you more confidence in assessing and building your collection.

WHAT TO COLLECT

The historical survey in Chapter 1 divided the period from the appearance of the watch to the present day into several separate fields. It is convenient to consider collecting in these same groups. The first, which ends in 1600, is probably best regarded as a museum field: a collection of these watches is not feasible, and the enthusiast will have to limit himself to seeing and, if possible, handling and photographing examples in museums and private collections. The second group, from the period 1600–75, are pre-balance spring watches, and again are largely museum pieces. This was a period of decoration, and the watches appeal to connoisseurs of enamelling and jewellery as well as to watch collectors; prices are relatively high. A 'puritan' watch in a plain oval case of blackened, tarnished silver, probably rather battered in appearance, might possibly be found, but with television, newspapers and books giving wide publicity to all antiques it is becoming less and less likely. The very scarcity of watches of this age means that the majority of those surviving are already in museums or private collections.

The next period offers more hope. After the introduction of the balance spring in 1675 there was a period of experiment, followed by 50 years of attractive, conservative design with decorated movements, champlevé dial etc. These watches are still expensive, but after 1750 plain white enamel-dialled watches in plain silver pair cases become quite common. Their prices require considerable thought before acceptance, but are within the range of credibility. Thus collecting becomes viable from about 1750. The next group (1775–1830) offers more possibilities, with a reasonable chance of a bargain. This is because the pair case disappears or becomes much less common towards the end of this period, and quite interesting watches are contained in unimpressive, single cases. Similarly, watches of the next two periods, 1830–1900 and 1900 onwards, may be found at prices to suit many collectors.

Two views of an English watch with cylinder escapement by Thomas Farr, Bristol, no 5316, in a silver case hallmarked 1815. The fusee has maintaining power, the balance is oversprung and a stop mechanism operates a bar which contacts the balance rim (missing). The movement is fitted with a dust cap.

There are, of course, expensive watches in these periods – quality must influence the price. After all, a brand-new wrist watch today can cost £10 or £10,000.

PRICES

Prices are difficult to discuss except in the context of the amount of money that the collector is prepared to spend. Even in the groups that have been nonchalantly bypassed as museum or connoisseur fields, there are a few people who will pay the enormous prices required. However, these are not ordinary collectors but investor collectors, who will seek expert advice from several sources before making a purchase.

Such prudence is also necessary in the more humble price ranges inhabited by the ordinary collector. Unless a watch has such special significance that the cost becomes less relevant, the collector must carefully assess its value both on the open market and as an asset to his collection, as suggested earlier.

In general, inflation increases prices, but there are also times when prices fall. This is sometimes associated with the 'state of the economy' or the 'price of bullion'. Unless it is your profession,

A chaffcutter (Debaufre type) escapement watch movement by Josh. Dumbell, Liverpool, no 817. The movement is in silver pair cases hallmarked 1825. The style shown by this view is closer to that of the lever, but from the look of the watch, with a contrate wheel, it could be taken as a verge watch. The double saw-tooth escape wheel operates on a single pallet on the balance staff; there is a fusee without maintaining power but no dust cap.

collecting is a luxury and a knowledge of price trends is useful.
This sort of information can only be acquired by regular visits to
watch fairs, antique shops, watch dealers' shops, auctions, car
boot sales and so on.

Generally, pair-case watches command a higher price than do
single-case watches. They are usually older, and are easily recognised as 'out of the ordinary' by the most uninformed owner;
indeed, they are often overpriced because of this. Similarly, any
form of decoration will increase the price, since it is easily recognised by the layman. Silver cases are less expensive than gold
ones, which are often completely outside the price range of the
ordinary buyer.

Two views of an English watch with duplex escapement by Wm Thomas, Strand, in
a silver hunter case hallmarked 1817. The fusee has maintaining power; the balance
is oversprung and has a brass rim with three timing screws. The notched cylinder is
of ruby. The movement has a dust cap.

If a watch comes from an eminent maker, its price can be many
times that of a similar item from a humbler craftsman. Thus the
ordinary collector is unlikely to purchase watches by the greatest
names, but he should be aware of perhaps fifty good makers so
that he can recognise a bargain when confronted with one.
There is not always time to go away and think about it when you

discover a Graham watch – but you must be sure that it is the right Graham and not a fake (unless the price is the correct one for an ordinary watchmaker's product).

COLLECTING 1775–1999 WATCHES

Rather than embarking on haphazard, magpie-like collecting, it is worth considering the possibilities within the 1775–1999 period in more detail.

1775–1830

The 1775–1830 group enables the collector to acquire examples of each of the escapements described in Chapter 2 (except the mass-produced pin levers). These escapements represent the majority of those produced in any quantity in the life history of the watch. There are other, rare escapements not discussed in Chapter 2, but these will be expensive. Of those listed, the verge and the table-roller English lever would be the easiest to find, but the remainder – with the exception of the detent – are obtainable at a reasonable price, given patience and a little luck.

Provided the collector knows in what sort of case he is likely to find the particular escapement he seeks, he may well be able to find what he wants and buy reasonably. However, if he asks for a specific, uncommon escapement he is bound to pay the prevailing price. To some dealers, a watch is a watch unless it is in a pair case, when it is a verge – a collector with some expertise is in an advantageous position on some occasions.

The most expensive watch in the common escapements of this period is the detent found in the pocket chronometer. This is likely to be outside the credible price range, but it might be possible to strike a collecting compromise and have a late nineteenth-century example, which will be more reasonably priced.

The 1775–1830 group is at the expensive end of the range of viable prices. However, it is one of the more interesting because of the considerable variety available. Probably the best way of approaching the period is to start at the 1830 end and work backwards as experience and confidence increase, or when a suitably priced earlier watch becomes available. It is important not to be over-confident, and it is probably wisest to study the various types of watch in museums, collections and illustrated literature before purchasing one. The table overleaf has been included as an aid to this study. It can be seen that the style of a case may give an idea of what is to be expected inside, and when the case is opened the style of the cock may give an idea of the type of escapement to be expected. Similarly, it is advantageous to know the style of letters used in the hallmarks of the period at the London, Birmingham and Chester assay offices. The majority of watches of this period that are available will be English and will have been marked at these centres. Continental watches will not necessarily carry a hallmark, but they are often identified by the bridge style of cock. If the watch has a maker's name and place of origin, then it may be possible to find the maker listed in Baillie's *Watchmakers and Clockmakers of the World* (see Bibliography). This period is about at the limit of usefulness of this book, for the comprehensive listings stop at 1825 and only makers of particular significance are listed for later years. There is now a second volume with the same title, compiled by B. Loomes (see Bibliography), which covers the period 1825–1880.

Two views of an early twentieth-century pocket chronometer, no 73602, by Maurice Dreyfus, Chaux de Fonds. The threequarter-plate watch has a helical balance spring, a pivoted detent escapement, a going barrel and a cut-compensated balance. The hunter case is engine-tuned all over and the hands are adjusted by a rocking bar, operated by a lever adjacent to the four-hour mark.

Watches 1775–1830

Escapement	Pair case	Single case	Cock	Recognition by	Photograph on page
Verge	yes	rarer	round and pierced	contrate wheel	20, 21, 139
Cylinder	yes	after 1810	round and pierced, solid later	escape wheel teeth	111
Chaffcutter	yes	yes	round and pierced, becoming lever shaped	contrate wheel	112
Rack lever	yes	rarer	special shape with words such as 'patent'	shape of lever	123
Massey lever	no	yes	lever shape with words such as 'patent'	balance-staff roller	126
Savage lever	no	yes	bulbous lever shape	pins on balance-staff roller	130
Duplex	yes	after 1810	bulbous lever shape	escape-wheel teeth (single beat)	113
Detent (in pocket chrono-meter)	yes	after 1800	lever shape	balance, helical spring (single beat)	–
English lever	no	yes	lever shape with words such as 'detached'	balance-staff roller	title page, 145

Common movement faults in this period
Balance-staff top pivot broken, fusee chain broken, mainspring broken, fusee drive pawls worn and slipping, balance-spring damage, watch out of beat, wear in balance jewels.

Left to right A late nineteenth-century watch, a barred movement from such a watch, and an early twentieth-century watch. Both watches are Swiss in origin and have mass-produced, going barrel-cylinder escapement movements. Such watches were made for a variety of markets and price ranges, with metal, silver, gold or enamelled cases and plain or decorated enamelled, silver or gold dials. The numerals on the dial of the right-hand watch show it was made for the Turkish market.

1830–1900

The 1830–1900 group will also offer a variety of escapements, but only in the first few years of the period. After 1840, choice practically ceases. The possibilities of collection lie therefore in the variation in form and layout of the classic English lever, the development of winding varieties (although the early designs are rare), balance types – brass, gold, steel, compensated etc – Swiss machine-made watches from 1840, variations in lever design, early pin-lever watches, and so on. Each of these groups, with the exception of the winding types, will offer a choice of watches at the lower end of the price range.

One particular type of watch of this period is the Swiss, ladies or 'fob' watch with a cylinder or lever escapement in a bar movement. These often have attractive, multicoloured dials. The watches appear in both gold and silver cases, and since the gold cases are often of low carat value and very thin material, they are not necessarily overpriced when considered as jewellery. Horologically, they are not often of great technical interest, but provided they are in working order they are collectable. Even if they need repair, it should be possible to establish the cost of the work and make a purchase at a suitable price. This remark is, of course, valid for any watch needing repair.

American watches of this period are another interesting possibility for the collector. It is unlikely that any of the scarce watches of pre-1850 will be found, for these early American watches were made in such small numbers. There is the possibility of sub-specialisation here within one company or in the series of companies which form the history of one company. A study of the American

US watches. *Top left* Waltham c.1865; *top right* Elgin c.1880; *centre left* Rockford c.1876; *centre right* Waltham c.1918; *bottom left* Hampden c.1882; *bottom right* Waltham Sapphire c.1927.

watches for sale in the UK shows that there are plenty of the Waltham group available in the 100-year span of the complex of companies. A disadvantage of this field to the British collector is that research would be largely concentrated in America, so that lengthy correspondence may be involved.

A possibility of collection over the span from 1775–1900 is to collect from one town, one maker or one year, but this will require a great deal of patience and searching. It might also be tempting to buy watches which fit the collection even though

Watches 1830–1999

Escapement	Pair case	Single case	Cock	Recognition by	Plate on page
Verge	rare	yes	simple	contrate	52
Cylinder (to c.1940)	no	yes	various	escape wheel	53
*Duplex Note the Waterbury (US)	rare	yes	various	tick	54
*Detent	rare	yes	plain	single beat	55
*English lever	rare	yes	becomes plainer	escape wheel	63
*Swiss lever	rare	yes	various	escape wheel	64
Pin lever	no	yes	various	escape wheel	65
Electric	no	yes	–	'quartz'	80
Electronic	no	yes	–	LED/LCD display	79

*implies full-plate or threequarter-plate models

Notes
Wrist watches have cylinder or lever escapements, a few have detent. Main case styles are single case with open face, half hunter or (full) hunter. There are various styles of open-face cases.

Common movement faults in this period
These are similar to those in the table for 1775–1830 on page 116. Faults in electric or electronic watches need specialist attention for both diagnosis and repair.

Footnote
Many Swiss movements have an 'extra' set of initials on one of the plates, often 'D.F. & C'. This is believed to be the name of the export/import agent Dimier Freres & Cie of Geneva.

they are overpriced or damaged beyond repair. This approach
would need good decisions.

1900 onwards

From 1900 onwards a collector might specialise in early wrist watches. Attention could be paid to the development of the less well-
known companies and the watches they produced. This should
again be possible without going to enormous expense. Alternatively,
a more expensive attempt could be made to acquire examples of the
variety of complex mechanisms that are available – repeating, chim-
ing, moonwork, date and day, chronograph, stop etc.

There are large numbers of watches available in this period
covering a wide range of prices, and it should be possible to build
an interesting collection without difficulty. Many of the watches
will be obtainable at prices which can be regarded as low enough
to provide essential experience, even if they prove to be not so
desirable as that experience accrues.

From an 'English' viewpoint, it would be possible to collect

UK watches. *Top row* Kelton; Kelton movement; first UK Timex with Timex movement (Dundee). *Bottom row* Smith's (Ystradgynlais); Newmark (Croydon); Westclox
(Strathleven).

watches made by the companies discussed in Chapter 3: the English Watch Company, William Ehrhardt, Rotherham, the Lancashire Watch Company, J. W. Benson and H. Williamson. Such a collection could be expanded to include the post-World War II essays of Smith/Ingersoll, Newmark, Timex UK and Westclox (Scotland).

Swiss watches. *Top left* Hebdomas eight-day watch; *top right* Arogno 294 digital watch; *Centreline top* Movado, hallmark 1917; *centre left* A.S.; *centre right* Maker unknown; *centreline bottom* A.S. 153; *bottom left* A.S. 427; *bottom right* A.S. 374. (A.S. is the movement maker A. Schild of Grenchen.)

Electric and electronic watches

Clearly, early electric and electronic watches are collector's items. There are considerable disadvantages in a collection of such watches, however, for the repair or replacement of parts presents problems which may be solved only by expert attention. One possibility for collecting such watches is to accept the non-functioning situation and to collect significant examples of this type of watch. It might be that if enough interest is shown in such watches, suitable replacement movements or parts will be made available by experts in this field.

One example of sufficient interest might be the Swatch, for which there is a worldwide club. Most Swatch catalogues give details of how to contact this organisation.

Military watches are collectable; the movement is Swiss. *Right* Irish watches are not common – John Donegan of Dublin is the best-known maker. The movement and case are numbered 13954 and the Dublin hallmark on the case (stamped JD) is for 1869.

A personal conclusion to this analysis of periods is that the most interesting approach would be to start collecting watches based on the year 1830 and to work concurrently backwards and forwards in time as suitable examples become available at the right price. This offers a wide scope and should eventually lead both to an interesting collection, and to the discovery of the collector's special interests. If the guidelines for purchasing have been observed, the resale or exchange of 'redundant' specimens may not be too costly, especially as it is likely that the time taken to acquire the collection will be spread over a long period.

The majority of the photographs in this book illustrate those watches made between 1775 and 1950. These photographs are not of watches by famous makers which are kept in museums, or in safes or bank vaults, but of ordinary watches belonging to ordinary people who have allowed their possessions to be photographed. They have almost all been purchased after 1970 and therefore represent what the ordinary collector can hope to find. There are other books available which show watches by famous makers with significant mechanical or horological features, or watches in superb cases of gold or enamel.

Two views of rack lever watches. *Left* This movement is by Thomas Savage, London, no 36806. It has adjustable bearings for the lever. The escape wheel is of steel and has 15 teeth. The cock is marked 'patent' and the 'Liverpool jewelling' is clearly seen. The watch has a stop piece, fusee with maintaining power, and a dust cap. The cock shape is typical of a rack lever watch. *Right* This watch is by Rt Roskell, Liverpool, no 7550, in plain silver pair cases hallmarked 1812 or 1832. It has adjustable bearings for the lever and a large escape wheel of brass with 30 teeth, as the watch has a three-wheel train with fusee. The cock is similar to that in the watch on the left. There is no jewelling and no dust cap.

AUTHENTICITY

The problem of fakes may be disturbing. It is perhaps relatively unimportant in watches costing very little, but there comes a price when faking is worthwhile. Nobody is going to make an 'antique' verge watch in the 1800 style, because the cost of making it will exceed the sale price unless the watch is inscribed with a well-known maker's name. In this case, the purchaser is likely to make sensible checks to establish the genuine age and not make an error. However, movements, dials and cases individually worth very little can be brought together and married to produce a watch which may be sold at an inflated price. A purchaser would do well to satisfy himself that all the parts belong together.

This problem should be alleviated by the assessment routine (see page 108), but there are watches that have been provided with new cases specially made for them: the watch shown on page 28 is an example. It is not easy to value such a watch, but a verge movement with bell repeating is not common, and someone obviously thought it worth replacing the case in 1843.

If buying from a watch dealer, it should be possible to get a written statement about the watch. Buying from a specialist is bound to cost more, for he has had to seek out the watch, possibly have it

Swiss watches. *Top left* Omega; *top right* Maker unknown, hallmark 1917; *centre* ETA 730, hallmark 1935; *bottom left* MST, hallmark 1911; *bottom right* ETA 128, hallmark 1922.

put into good order, and hold stock, which will tie up capital; but to purchase from the professional may be a safeguard against false purchases in the more expensive ranges. Another source of watches is by exchange, both with dealers and friends. As a collection grows there are often too many of one type of watch, or your field of interest shifts.

Movements are sometimes modernised. A duplex of 1810 may have broken in 1860 and the owner may have had the escapement changed to a lever, which would be considerably easier to get repaired and more satisfactory as a watch. This work is perfectly genuine and may be worth collecting; the error here would be to assume the watch to be an early lever.

This Swiss watch is in a blued-steel case. The dial crack is obvious and there is other damage around the month hand. The display gives time, day, date, month and moon state. Other, more expensive watches might include chronograph and repeating mechanisms.

However much care is taken, many collectors will find that some of their watches are not exactly what they imagined them to be at the time of purchase. This is not to suggest that they have been fooled, or that there is a huge industry of putting together parts to sell, but that repairs and replacements made in good faith some considerable time ago by watchmakers are not easy to detect (the watchmaker who recased or modernised a watch in a crude fashion would not have attracted a lot of custom). A watch that is 200 years old would have to have led a very sheltered life to have survived without some repair.

INSURANCE

A collection of watches will represent an increasing investment as time passes (not necessarily by inflation but merely by increasing numbers), and it is worth looking at the relevant clauses in your household insurance policy. The collection may need special mention to be covered, especially if arranged in some sort of display. Accidental damage due to dropping or other causes should also be considered – repairs may be expensive.

COLLECTING WATCH PARTS AND ACCESSORIES

For some collectors, the acquisition of complete watches is unnecessary. An interest in the mechanical form of the watch rather than the complete instrument is a perfectly good reason for collecting, and in this case the purchase of a quantity of gold or silver which hides the part of interest seems pointless. In the

Two views of a Massey lever watch in a silver case hallmarked 1828. The maker may be T. Lamb, but the style of the engraving suggests an owner and date. The roller on the balance staff is of the second Massey type, with a jewel between two supports. The watch has 'Liverpool jewelling', fusee with maintaining power, a stop piece and a dust cap. The cock is marked 'patent'.

past, many movements have been removed (gently or roughly) from their precious-metal cases; the cases have then been melted down, and movements are often available at a fraction of the cost of the complete watch. The older the movements, the rarer they become and the more decoration they exhibit. To many collectors, the movement is actually more attractive than the watch, irrespective of the mechanical interest. Sometimes in the past, the only part of the movement retained in the scrapping process has been the balance cock. Cocks have even been 'faked' and used as jewellery. There is no excuse for anyone breaking up watches today, but unfortunately the practice continues and pocket-watch movements from single cases are still becoming available. Similarly, wrist-watch movements can be acquired, and may be used as part of a collection. Dials could be collected, or hands, but these are less likely to be found as separate items.

Watch keys may be collected but this is a very limited field, for the early and interesting varieties are rare. The individual key, perhaps enamelled in the same style as the watch, should of course remain with the watch, but it is sometimes possible to find a lone survivor.

Watch papers are sometimes found in the back of a case. They are advertising material, but on the back they often have a description of a repair and its cost with a date.

Watch papers are interesting in their own right. These are small, roughly cut, circular pieces of paper inserted between the pair cases (or sometimes in the back of a double-bottom case), usually with the maker's, repairer's or seller's name and place of work printed on it, but often containing other information. They are not common, for they are easily lost or the original owner may have discarded the 'advertising'.

Watch chains with hanging ephemera (chatelaines) are sometimes found, but are not common enough to enable a collection really to be viable. Seals on chains have their own following and are hardly vital to a watch, although they could possibly provide some evidence of the history of a particular watch.

COLLECTING INFORMATION

The data available on every watch, movement and accessory in a collection should be recorded and for this a card-index system is useful. The watch or movement should be studied, researched and written up because in this way, the collector learns more about watches and will be led to new and perhaps more interesting discoveries. Only by drawing on the store of information in books, museums and collections can the collector become more knowledgeable himself, and only he can correlate the information in the form relevant to his own collection.

A watch acquired quite casually may have an interesting and traceable history or its maker may have unknown connections.

Watch stands of all types are collectable. The example shown is silver (hallmark 1906) and was probably sold complete with the Japy watch. Many stands are made of wood and may be used to display a collection.

Most watches will have no traceable history or yield no new information, but it is worth making some effort to follow up any promising lead. The card index should also record mechanical details, work done etc, so that nothing is left to memory. Even the most simple work done is often forgotten, and it may prove important later when more information or experience suggests further work on the watch.

Watch literature is important. Books and journals that are contemporary with the period of collection should be bought or borrowed, which is feasible enough from perhaps 1850 onwards, with some earlier sources also being available. The use of libraries (public, reference, university, museum and institution) is essential and library research facilities such as copies of old documents, out-of-print books etc are usually available to the genuine enquirer.

Reading about the methods used by old watchmakers helps develop an understanding of old watches. Similarly, the contemporary difficulties of timekeeping, regulation, balance springs and balances may be better understood if they are studied in the original documents. The mathematically minded collector could reach a better understanding of the friction, forces and vibrations in a watch, that will explain the reasons for the success or failure of many innovations.

Photography might be used to 'collect' items normally unavailable from a certain field. In particular this applies to obtaining permission to photograph watches in museums and private collections. In less elevated watch fields, it might be possible to photograph quite ordinary watches in friends' collections or dealers' stock. I use a 35mm SLR body and a lens which will produce full-size images using standard processing. A collection representative of a period can consist of both watches and photographs.

MANUFACTURING WATCHES

Manufacturing a watch is perhaps the ultimate achievement for a mechanically minded collector, involving as it does the deployment of a variety of skills, such as machining, fitting, gilding, casemaking and enamelling. It may only be suited to a collector possessing a workshop and sufficient skill to justify the time involved, but the techniques and some of the equipment may be obtained and used at evening classes in local schools of arts and crafts. The tools may be acquired as a separate collection, for many of them are specialised and obsolete; the repairer does not need them for modern factory-made watches.

In all the mass of suggestion and discussion of research, it is important that the collector is not deterred. It is not necessary to seek these specialisations, for there is also a great deal of pleasure to be had in finding and keeping a haphazard collection of watches simply because you like them. This is the most important aspect of collecting: the magpie instinct. Think of 'millenium watches' – would you like to collect them?

A Savage two-pin lever movement by Thos. Sherwood, Leeds, no 1234. The date is about 1820. The style of the balance is similar to that of a verge watch but it is over-sprung. The movement has a fusee with maintaining power and a dust cap.

CHAPTER 5

REPAIRING

A watch may be found to be in any of several states. If it is not working, due to damage such as may be caused by bending or breakage of parts, missing parts, corrosion etc, its value both in terms of money and as part of a collection may be considerably reduced. However, it may be so old or so rare that it has more value in an unrestored condition than a restored one. A watch may be in what can be called ticking order: complete and working but in a desultory and discontinuous way, a state possibly caused by excessive wear, dirt, or minor damage. A watch in working order may be expected to go continuously under reasonable conditions, but the timekeeping may be poor and unreliable. The watch may also be reluctant to go satisfactorily in all positions, working perfectly well on a desk when the dial is up, but soon stopping when carried about in the pocket. Finally, a watch may be in good working order, or as good as the average watch in daily use; it can be used to tell the time as accurately as its inherent capabilities could ever allow. It is not realistic to expect a verge watch to give the same performance as a late Victorian lever watch. It is also unrealistic to expect a watch to give of its best unless it regularly receives expert attention.

A collector must decide whether he requires all the watches in his collection to work, or whether they can remain as he finds them. There will be protagonists for both sides, but it seems reasonable that, with the exception of those mentioned above as being so rare that restoration could be detrimental, a watch which is a device for telling the time should be able to work in some fashion. Assuming that the collector subscribes to this view, he may take his watches to a watchmaker, but this often means waiting a long time, for the work may involve making a missing part, with time-consuming trial-and-error fitting. Also, a watchmaker is usually so busy with the regular maintenance and repair of modern watches that he has little time for investigating and repairing an antique watch. The repair could be very expensive and may exceed the value of the watch to the collector. Therefore, when a damaged watch does become available, the collector employing a watchmaker to do his repairs would be wise to obtain an estimate of the cost of restoration before he decides on purchase.

Many collectors choose to repair their watches themselves. This can be a much more interesting proposition, as the work is an extremely time-consuming and absorbing hobby, bringing considerable satisfaction when a watch which may have been broken for a long period is finally brought to life. There are two distinct approaches: either as an amateur, or as a spare-time watchmaker. The latter approach will mean tackling the work in a professional way, purchasing all the equipment needed to restore or renew any part of any watch. The collector taking this approach may already have the necessary skills, but if not he must teach himself or attend classes in watchmaking, silversmithing, enamelling etc. He will also need to make a considerable investment in equipment. These subjects have their own literature. After some years of interest, and probably after some experience with the amateur approach discussed below, the collector is more likely to see himself as a spare-time watchmaker. Eventually the professional will seriously consider making his own watch, preferably incorporating as many complex mechanisms as he can.

The amateur approach involves starting with very limited aims, possibly twofold: firstly, to be able to restore the majority of one's watches to at least ticking, and preferably working, order so that they will run continuously in at least one position; and secondly, to gain skill, confidence and experience so that more ambitious work can be attempted. Initially an amateur will not be capable of making new parts, but he can use ingenuity and spare parts from other watches to solve problems. His lack of skill and equipment will mean that at first there will be watches that he will have to accept as not working, but as progress is made the problems that seemed insurmountable will become solvable. Because of their size, wrist watches would be a poor field for him to tackle, except that he can gain experience inexpensively and spare parts for such modern watches may still be available. One of the most important characteristics to develop is patience: many jobs take a very long time and the restoration of a badly damaged watch could spread over months.

The remainder of this chapter is concerned with this amateur approach; it suggests tools, materials and rules, and finally describes some basic operations. Some of the suggestions will cause head-shaking by professionals: however, it is clearly stated in the 'rules' that everything attempted should be capable of being undone without any permanent record appearing on the repaired watch. This means that no damage has been done and the work can be repeated in a more professional fashion when the skills have been learned. It should also be remembered that many of the watches are beyond economical repair by a watch-maker, so in fact the amateur is improving the watch as well as having the pleasure of the work.

Some rules for the amateur watch repairer

1 Never do anything permanent. The watch should always be left so that it could be returned to its original state. Some time later a better way of repair may occur, skills may improve, more equip-ment may become available or the watch may turn out to be more desirable in the original unrestored state, being discovered to be unique or rare.

2 Never start to dismantle a watch with the mainspring wound. This is particularly important, because the majority of escape-ments will allow the watch to run at high speed as the balance staff is withdrawn. If this happens there is a good chance of dam-age. The mainspring should therefore be let down.

3 Be sure always to study, and if necessary sketch or photograph, the original state before starting work. Never dismantle mecha-nisms that are not recognised and understood. Find out first what it is and how it works, then decide how to proceed.

4 Be careful of applying force unless the construction is well understood. Thus unscrewing a left-hand thread with the cus-tomary anticlockwise motion will in fact result in breaking the head from the screw – a simple example of misapplied force.

5 Always work on a clean bench, and put the microscopic parts away as they are removed.

6 Always make a new part rather than cut or drill an existing one.

Cutting and drilling are irreversible. Some modern glues are very strong and parts may be glued into place rather than screwed. However, some modern glues are too strong and may equally cause damage.

TOOLS AND MATERIALS

The lists below might be regarded as the minimum needed to make small repairs. They are not exhaustive, and each individual will add items as they become necessary. One of the most difficult aspects of working on watch parts is the small sizes involved, and tools and jigs will be required to enable some things to be done. These are devised as and when they are required, but it is useful to remember that all these jobs have been done before, and information in books or from watchmakers is invaluable. The illustrations found in books can also be very useful.

Tools

screwdrivers	pin vices
tweezers	knife
needle files	eyeglass
broaches (cutting and polishing)	Archimedean drill
boxwood movement stands	drills and pin chuck
small vice	micrometer
stake and punches	hammers
screwhead slotting file	dust covers
work bench with swivel light	spirit lamp

It should be noted that these are specialist tools which are very small and may not be available from a local shop. Specialist journals should be consulted.

Materials

small-diameter brass and steel wire for pins etc
benzine and ammonia for cleaning
lubricating oil
releasing oil
replacement glasses
replacement mainsprings
replacement staffs

glues (epoxy resin, shellac)
pegwood
Arkansas stone
fine emery paper, polishing powder
old watch movements and parts from all sources

SPECIAL EQUIPMENT

You may also require some more specialised tools for undertaking particular tasks.

A *mainspring winder* is an essential device for winding a replacement mainspring into a tight coil to enable the spring to be put into the barrel with no axial distortion. This avoids problems with lack of power due to friction with the barrel lid, or in extreme cases, the axial force possibly displacing the lid.

A *mainspring punch* is a pair of 'pliers' where one side of the gripping surfaces has indentations to match the steel punches which protrude from the other gripping surface. Thus when the 'pliers' are closed, the appropriate punch makes the hole in the spring which will then engage on the stud on the inside of the barrel rim.

Watchmaker's turns are a small lathe held in a vice. The work to be turned is held between the 'runners' and is rotated by a bow operating a pulley of appropriate size pushed on to the work. (If the work is a part with a hole in the centre, it is put on to a small tapered bar with an integral pulley.) The turns have a fixed headstock and movable tailstock. Between these two is a movable tool rest to support the cutting graver. Right-handed users will use the left hand for the bow and hold the cutting tool against the rest with the right hand.

A *watchmaker's lathe* is a much-enhanced piece of equipment with multiple accessories. There is no reason why a collector should purchase one until he has decided whether he wishes to pursue this path in his horological interests.

A *staking tool* is used for the same tasks as the simple stake and a set of punches. The tool has a horizontal rotating stake to support the work. The appropriate punch (part of the staking tool kit) is held in a vertical guide which is situated over the chosen hole in the rotating stake. This avoids the problem of the punch slipping and damaging the watch part.

A *re-pivoting tool* is used to replace a broken pivot on a wheel arbor. (This work may also be performed in turns with a long hollow runner.) The slightly tempered arbor is first prepared for drilling by flattening the broken stub. It is then put into the tool with the broken pivot placed at the drill end of the U-shaped re-pivoting tool. The other end of the tool has a runner with a hole to support the good pivot. This runner has a pulley to be rotated by a bow. The pulley has two stiff wires which 'engage' with the wheel spokes to rotate the arbor which is to be drilled. Drilling is carried out carefully until the hole is deep enough to accept a new piece of pivot material. The new pivot is finished and polished using another special runner for support, or a special Jacot (polishing) tool.

The large lathe or mandrel may be used for watch plate work. Just below is a pair of turns, together with some pulleys to rotate the work with a bow. Three cutting tools, known as gravers, are also shown; these are made of hardened steel with a cutting end shaped by the user to suit a particular job. On the right, in a box, is a Jacot tool, which is used for polishing pivots. In front of the box, on the left, is a small 'turns' for drilling holes in arbors to replace a broken pivot. Below this is a hand-held vice and a screw cutting die for holes of various sizes. In front of the box, on the right, is a pitching or depthing tool for setting up wheel-and-pinion engagement and scribing lines on plates. Below the depthing tool is a mainspring winder (the top bar is missing).

A *depthing tool* is used to plant a new wheel in a train. It consists of a V-shaped bed which can be opened or closed to suit the diameters of the wheel and pinion whose engagement is being set correctly. The arbors are held in the two V-shaped ends of the tool by runners, which protrude through the V-ends and have hardened points. When the engagement of the wheel and pinion being adjusted is satisfactory, the pointed ends are used to scribe a line on the pillar plate under the dial. The process is repeated, with the wheel attached to the engaging pinion being tested with the next pinion in the train. A second line is scratched on the plate and the correct position for the arbor being planted is then drilled. This is not easy to describe, but if the surface of the plate underneath the dial of an English lever watch of the mid-nineteenth century is examined, it is usually possible to see scribed arcs upon it.

The descriptions given of all these specialised tools are hardly adequate unless you have the equipment in your possession. The photograph opposite shows all except the punch and the staking tool (other tools are also shown). Books such as De Carle's *Practical Watch Repairing* (see Bibliography) deal well with their use for lever-watch repairs. Work on watches with older escapements or on more modern watches will require other books.

WORKING WITH WATCHES

A description of the methods used to correct a number of faults follows. All the methods are suited to amateurs and should not be regarded as the only way of tackling a job. All have been used and have been successful in their limited aim.

Letting down a mainspring

In early verge watches the mainspring set up is by worm and wheel between the plates. A small key is required to fit the winding square on the end of the worm shaft, and the shaft is turned until there is no tension left in the chain connecting the fusee and spring barrel. If the watch is fully wound, this will require a considerable amount of turning.

In later verge watches and English watches of the nineteenth century the set up is under the dial. Remove the dial (see 'Stripping and reassembling a lever watch', page 144) and fit a

Fig 1 Set up.

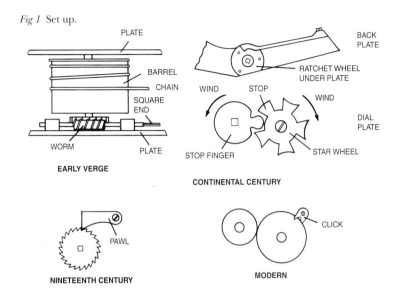

key into the square that protrudes from the end of the ratchet wheel. Slacken the screw which holds the pawl tight. By initially tightening the spring, the load can be taken from the pawl so that it can be pushed back with pegwood to free the rachet. Hold the key firmly so that the spring does not fly undone. If there is any danger of slipping, push the pawl back and perform the operation in increments. At the end of the letting down, remove the loose ratchet wheel. Sometimes the amount of square which protrudes through the ratchet wheel is too short for this method, in which case let the ratchet off, tooth by tooth, using a screwdriver as a detent when the pawl is out of engagement.

In Continental key-wind watches with a going barrel and barred movement there will be two places for attention when letting down the mainspring. Viewed from the barred side, the click keeping the spring wound is close to the winding square, forming a V-shaped detent integral with, and on the end of, a leaf spring. This detent can be levered back and the spring allowed to unwind with the key held in the hand. Under the dial there will be a star wheel, which is the stopwork fitted to limit the use of the spring to the middle portion only. Hold the lower (adjacent) square by the key and remove the star wheel. Holding the key in the hand, allow the set up to unwind. Examine the action of the

star wheel before removing it, so that later it can be reset correctly with about threequarters of a turn of set up. It can be seen
that one arm of the star has a convex end rather than a concave
one and will not pass the stop. The number of complete rotations
of the winding barrel is limited by the number of concave ends.

This group of four photographs shows the development of the verge watch movement from 1700 to 1850. All the watches have a fusee and chain. *Top left* This is a
movement by Cornelius Manley of Norwich, c.1700, and a large pierced cock with
D-shaped foot and ears. It has Egyptian pillars and Tompion-style regulation with
figure plate. *Top right* This movement by Wm Ransom of London is c.1760 and
shows a smaller pierced cock with a smaller fan-shaped foot. The movement has
square pillars and is thinner than that on the left. *Bottom left* This movement by Jas
Dysart of London, c.1820, is the largest movement shown. It has a solid foot to the
cock and later Bosley-style regulation. The pillars are cylindrical. *Bottom right* The
smallest movement, c.1845, has a bridge over the barrel for easy removal of the
mainspring. The style of the cock resembles that of a lever watch of the period.

In modern watches and wrist watches, the spring click holding the winding train can easily be seen above the top plate. The winding button is tensioned in the wind direction to remove the load from the click, which is eased out of engagement with peg-wood. The winding button must be held during this operation and then allowed to rotate slowly as the spring unwinds. It may be necessary to allow the click to re-engage, and then to perform the letting down in increments.

There are other, less common mainspring arrangements, and careful attention to rule 3 in the list on page 133 is advised so that accidents are avoided.

Putting into beat

Often a watch will tick unevenly and stop because the neutral mid-position of the balance vibration is not in line with the correct point of the engaging lever, crown wheel or escape wheel. The watch is *out of beat*. To set the watch *in beat*, these points must be lined up correctly. A study of the action of the various escapements in Chapter 2 will indicate the correct position. For example, the mid-vibration position for the balance of a table-roller lever watch is with the impulse jewel in the centre of the lever notch. Since the lever does not remain in this position but is to one side or the other, some judgement is required. This is true of all levers; and special care will be needed with a rack lever, since there is positive engagement between the balance staff and the lever.

The first step in adjusting the beat is to let down the mainspring. This has been described above and ensures that no accidents can occur. The neutral position of the balance can now be altered by unpinning the outer end of the balance spring from the post on the plate and rotating the balance the required amount in the appropriate direction. It may be necessary to remove the cock to gain access to the pin securing the outer end of the spring. If there is not enough spring available, or it has been broken off, then adjustment may be gained at the other end of the spring. Remove the balance and rotate the split collet holding the inner end of the spring to provide adjustment at the outer end. The collet is a friction fit on the balance staff. Should this action be necessary, it is likely that the balance spring has been broken off short, and this means that once the watch is in beat it will gain. When making adjustments to balance springs extreme care must be taken, for their shape is important in maintaining isochronous performance. Once they are bent out of shape, the rate of the watch will vary, especially if used in a variety of positions. (See also 'Balance springs' below.)

Balance springs

If a balance spring is missing or damaged it should be replaced with one of the appropriate strength and size. These may be obtained from a watch materials supplier or from an old watch. The count of the train (determined as indicated below) will be required to give the strength, and the diameter must fit the regulator curb pins. After the new spring has been fitted to the collet in the correct manner as shown in Fig 2, make sure that it is central and in the correct plane. The holding pin should have a flat side against the spring and should not protrude beyond the hole through the collet. Place the collet on the balance staff in such a position that the watch can be put into beat.

Next, test the rate of the new spring. Grip the spring with a pair of tweezers at the point where the curb pins will hold, and rest the pivot of the balance on a smooth surface such as a watch glass. Set the balance vibrating and count the vibrations. Minor adjustments are possible by the putting-into-beat methods described above, but if the count is obviously wrong then a different spring should be tried. The most likely counts to be found

Fig 2 Balance spring.

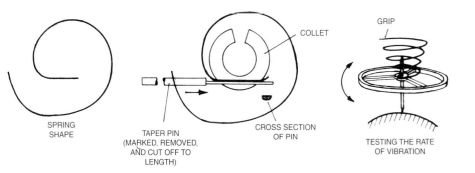

SPRING
SHAPE

TAPER PIN
(MARKED, REMOVED,
AND CUT OFF TO
LENGTH)

COLLET

CROSS SECTION
OF PIN

GRIP

TESTING THE RATE
OF VIBRATION

when adapting springs are 14,400 or 18,000 per hour, between which there should be no confusion. If a spring cannot be found to give the correct rate, one that is too stiff should be chosen and the height carefully and evenly reduced by rubbing on ground glass covered with grinding paste. The spring should be held embedded in a cork.

Balance springs are difficult to adjust accurately to suit all positions and temperatures, and the amateur is unlikely to achieve perfect timekeeping without considerable experience. If the watch was not working but does after attention to the balance spring, then a big step has been taken, and it is unwise to continue to seek perfection unless this is your particular interest.

Stripping and reassembling a verge watch

The movement must first be removed from the case by pushing out the hinge pin. This may be tapered, and care should be taken to push in the right direction. If it is stuck, apply a little releasing oil, but it is not worth using undue force as the movement can be removed without disturbing this pin. To do this, remove the hands and the dial (see below). The effect of removing the dial will be to release the movement from the case, but leaving the dial attached. The remainder of the operations are presented as a list. These instructions are for the older verge watch with complete unbroken top, and pillar plates with worm set up between them. Later verge watches with a bar on the pillar plate, separate barrel cover on the top plate, and rachet-and-pawl set up are stripped by the lever method (see page 144).

Before starting to strip, make sure that reassembly is possible. Put all the parts away as they are removed, and as you do so, label and make sketches as necessary. Try to avoid having the parts of several different watches on the same bench.

1 Remove the hands using homemade levers or a puller. Protect the dial when using levers.
2 Remove the pins holding the dial or dial backing ring from the pillar plate. There will be three feet protruding through the plate which are secured by pins. Beware of the motion work, which may fall loose and get lost as the dial is removed. Lift the dial upwards off the movement to avoid this problem.

3 Remove the motion work.

4 Let down the mainspring. Support the movement on a box-wood stand while doing this.

5 Remove the balance cock, which is secured by a single screw.

6 Remove the taper pin holding the balance spring to the block on the plate. Remove the balance-spring end from the block and the curb pins on the regulator.

7 Carefully lift out the balance complete with spring.

8 Remove the regulator mechanism and other decorative pieces from the plate.

9 Push out the four pins holding the plate to the pillars, noting the position of the short pins.

10 Carefully lift off the top plate. The crown wheel will remain attached to the plate.

11 Remove wheels, barrel, fusee chain and fusee. The centre wheel will remain attached to the pillar plate.

12 If required, remove the crown wheel by pulling out the end bearing which is a taper fit.

13 If required, remove the centre wheel by removing the cannon pinion from the dial side of the pillar plate.

14 Check the action of the fusee stop piece (see page 153).

Reassembly is achieved by reversing the sequence of operations. The watch is built up on a boxwood stand, and the difficult step is replacing the top plate, which must pass over the pivots of no less than five pieces; this must be done very carefully to avoid breaking a pivot. After the top plate has been replaced, assembly is straightforward. Lubricate the pivots of the watch but not the wheel teeth. Fit the balance so that the watch will be in beat. Finally test the running by applying a slight clockwise twisting force to the winding square of the fusee, which simulates the mainspring action.

The last part to be fitted to the movement is the fusee chain. Wind this on to the fusee by the normal winding mechanism, and insert the hook at the end into the slit in the barrel (the barrel hook is the barbed one). Turn the set-up worm to apply tension to the chain, and allow the watch to run until this set up is used up. Apply more set up and again allow the watch to run. Eventually the fusee chain will be on the barrel with the spring unwound. Finally, replace the motion work, dial and hands.

Stripping and reassembling a lever watch

Remove the movement from the case by pushing out the hinge pin in a similar way to that described for the verge watch. Some late levers may be held in place with screws with either half heads to be rotated free or whole heads to be removed completely. These types of movement pass out through the front of the case. If a button wind is fitted, it must either be pulled back out of engagement or the small screw holding the stem in place slackened to allow the winding stem to be removed. The instructions below apply to the nineteenth-century lever watch, but with obvious adaptations they also apply to the late verge watch and to other nineteenth-century English watches with a variety of escapements.

1–7 Remove the hands, dial and motion work. Let down the mainspring. Remove the cock, the taper pin holding the balance spring and the balance. For detailed instructions for these operations see steps 1–7 on pages 142–3. The watch should be on a boxwood stand.

8 Remove the barrel covering plate held by two screws. Never attempt to do this with the balance in place or with the mainspring wound.

9 Remove the barrel and the fusee chain if on the barrel. If the fusee chain hook is stuck in the fusee, a little releasing oil may help. The set up ratchet wheel from the back of the plate should have been removed when the mainspring was let down. If it was not, it will now be loose.

10 Turn the watch over and remove the bar across the third wheel. This is held by two screws. Remove the third wheel.

11 Turn the watch over and remove the four pins holding the top plate to the pillars. Note where the short pins belong.

12 Turn the watch over. Carefully separate the plates, leaving the wheels on the top plate on the stand. The centre wheel will come with the pillar plate.

13 Remove the wheels, fusee (and chain) and the maintaining-power detent. Remove the lever, which is partly underneath the balance bottom bearing.

14 If required, remove the centre wheel by removing the push-fit cannon pinion from the dial side of the pillar plate.

15 Check the action of the fusee stop piece (see page 153).

To reassemble, reverse the operations. Place the top plate on the stand, insert the lever under the balance bottom bearing and between the banking pins, and then insert the escape wheel, fourth wheel, fusee and the fusee detent. The pillar plate with centre wheel is placed on top, and the only pivots to be fitted are the lever, the escape wheel and the fusee detent. Great care is still

English lever movements of the period 1830–60. All the watches have a fusee with maintaining power. *Top left* A table-roller lever of about 1830 with deeply chiselled cock inscribed 'detach'd'. Maker unknown. It has 'Liverpool jewelling' and a big balance wheel overlapping the bridge over the barrel. *Top right* A decorated movement of about 1840 by Wm Roberts, Liverpool, which has the spring set up on the plate rather than under the dial. The arrangement of jewelling is further anticlockwise than conventional because the escape wheel, fourth and third wheels are to the right of the lever (viewed from the top plate) rather than to the left. *Bottom left* A threequarter-plate movement by Josh. Penlington, Liverpool, c.1830, has auxiliary compensation on the balance to counteract middle-temperature error. It is marked 'detach'd lever'. *Bottom right* A movement by Josh. Blanchard, Preston, of the 1860 period, when the style of the lever watch is fully developed. Of good quality with compensation balance, and fully jewelled except for the lever pivots. The lever is curve-sided whereas the three earlier movements have straight-sided levers. The cock has minimal decoration.

needed to avoid breaking a pivot. When inserted, turn the movement over on the stand (being careful to hold it together) and
insert the four pins holding the top plate to the pillars. Again,
reverse the movement on the stand, replace the fourth wheel in
its pivot as it will have fallen out with the reversals, insert the
third wheel, and fit the bar across the plate supporting the fusee
and the third and fourth wheel pivots. Reverse the movement on
the stand and replace the barrel and barrel plate. Lubricate the
pivots of the watch but not the wheel teeth.

The chain of a lever watch can be put back before the balance
is fitted – this avoids damaging the balance. However, the chain
cannot be put back at this stage in any other escapement. In the
lever watch, the chain is wound on to the barrel using the square
protruding from the dial plate underside: hold the movement so
that the barrel arbor is horizontal and allow the chain to run
under a finger, which keeps a small pull on the chain. If the
chain is slack, it may slip off the barrel and jam on the arbor. If
this happens, remove the barrel plate and barrel, then the chain,
and start again. Pulling at the chain may break it. When the
chain is on the barrel, continue to use the finger to hold it and
hook the free end to the fusee. (Or, if not successful, try a small

A view of a lever watch with fusee, showing the conventional layout of the train.
Balance and balance staff have been removed.

piece of sticky tape to hold the chain in place but *remember to remove it* as soon as the chain is tensioned.)

Now replace the ratchet wheel on to the square and apply set up to put the chain in tension. Push the pawl into the ratchet with pegwood, and tighten the pawl screw. Replace the balance and pin it so that the watch is in beat. Make sure that the impulse jewel is on the correct side of the lever jaws or the train will be locked. Replace the cock. The running can be tested immediately, as the chain is already in tension under the set up. Finally, replace the motion work, dial and hands.

Going-barrel (pocket) watches have no fusee and chain, and the centre wheel is driven by the going-barrel wheel (or a dummy wheel). The instructions for fusee and chain replacement are not relevant: the going-barrel wheel(s) is (are) simply replaced in the appropriate position, and set up is applied using the ratchet wheel under the pillar plate.

In the late verge watch, and other escapements except the lever, the balance and cock must be replaced before the chain is put in. In these cases, test the running by applying a small clockwise torque to the fusee winding square, which simulates the mainspring action. Then wind the chain on to the fusee using the normal winding mechanism, and gradually work it back on to the barrel by applying small amounts of set up. This technique is described on page 143. If the watch has a dust cap, this should be fitted when applying the small increments of set up; it will protect the balance.

Stripping and reassembling other escapements

Watches with escapements other than verge or lever will usually be of similar construction to the early verge or the lever, so the techniques already described may be used. The most important points are to make sure that the mainspring is let down before removing the balance, and not to apply any spring tension until the balance is replaced. Pocket chronometers, unrecognised escapements, tourbillons, karrusels etc should not be disturbed by the amateur.

There will also be watches that are assembled in different ways. The half-plate design is an obvious case. These may be taken apart carefully without damage, provided that the basic methods already described are used.

In key-wind watches with the set hands square on the pillar plate, both the centre wheel and the cannon pinion are push-fitted to the arbor. The arbor is removed complete with the set hands square from the pillar plate, after removing the hands, dial and cannon pinion.

Cleaning a watch

When a watch has been taken apart it may be cleaned by immersing it in benzine. However, the mainspring in the barrel, the balance and the lever should not be immersed but should be brushed clean with benzine; prolonged immersion could disturb jewel settings. Do not be tempted to use petrol or any other solvent. Benzine immersion degreases the parts, which should then be brushed clean and dry. Holes and jewels should be cleaned out with pieces of pegwood. Pinion and wheel teeth may also be pegged clean if covered with dirt. Benzine treatment should bring the whole watch to a satisfactory state, but if the gilding on plates and wheels is not good enough it may be cleaned with a weak solution of ammonia applied by a soft brush and washed off with water. Thorough drying is essential in order to avoid rust in the future. The ammonia treatment should be tested on a spare part from another watch in order to judge the strength and the result.

Pivots may be polished using a polishing powder in a hole in the end of a piece of pegwood. This method will not repair damage but merely buff up the pivot, which should afterwards be washed off in benzine; this is a treatment of dubious merit which could result in breaking a pivot, and in general, unless pivots show bad blemishes, they should be left alone.

After a watch has been cleaned, the parts should only be touched at their edges, or they should be held in tissue paper. This avoids transferring grease or perspiration to the parts. However, a fingerprint is preferable to a dropped watch, so use your discretion when handling them.

Lubrication

Only the pivots of a watch train and escapement need to be lubricated. The wheel teeth should not be treated with any oil. The oil should be applied with a fine wire dropper to the reservoirs in the plates (or, in early watches, to the pivot point). Only the smallest amount should be put on the dropper to avoid drips in the wrong place, besides which any excess oil will spread to the wrong places. Also, dust will stick to oil and form abrasive mixtures. A good-quality oil should be chosen, and the container should be kept tightly closed, as the atmosphere will have a harmful effect on it.

Strictly speaking, different grades of oil should be used on train and balance pivots, but for the collector whose watches are not going to run continuously a single grade may be chosen. Similarly, in a lever watch the pallets should have a microscopic amount of lubrication, but for the collector it may be better to keep oil away from this area. The watch is not going to run continuously and some modern watches do not use pallet lubrication at all. The impulse jewel in a lever watch should not be lubricated. Oil is, however, required in the spring barrel. The surface of the spring should be coated to avoid rust, and the top and bottom of the inner barrel surface on which the spring rubs needs lubrication.

Checking for faults

If a watch will not tick or show any signs of motion, and there is no obvious damage, it should be carefully stripped and cleaned. The train pivots should be examined for damage, and any really bad pivots straightened, cleaned, polished or renewed (see page 158). The holes in the plates should be pegged out and examined for damage. Really bad holes should then be cleaned up with a polishing broach, and extreme cases may be bushed. Unless accurate timekeeping is the aim, it should suffice to clean up the holes rather than rebush in cases of moderate wear. Bent or damaged wheels should be straightened or renewed (see page 158). Check the action of the fusee to see that the power is being transmitted. It has been assumed that a broken mainspring is not the cause of the trouble, but the barrel cover should be removed to examine the state of the spring, and to see that it is correctly attached at either end.

Reassemble the train without the fusee chain and without the escapement (balance or balance and lever), then apply a clockwise torque to the fusee winding square to simulate the mainspring action. The train should run freely. If it does not, there is an undetected fault. When satisfactory, fit the balance so that the watch is in beat; then again apply clockwise torque to the winding square. The watch should run. If not, there is a fault in the escapement to be detected.

In the case of the lever watch, the fusee chain may be replaced before the balance is fitted. It will have been necessary to take the train to pieces to insert the lever, but when the chain is tensioned the action of the lever may be examined. When the lever is given a small displacement from the banking pin (by pushing the *fork* end across) and then released, the draw should snap the lever back to the banking pin. When given a larger displacement in order to unlock the train, the lever should then snap over to the other banking pin. This action can be tested for a complete revolution of the escape wheel, that is, 30 escapes for a 15-tooth wheel. Provided this action is satisfactory, the watch should work once the balance is fitted.

Mainspring replacement

The mainspring may be inspected after the barrel has been removed from the watch. To do this, prise off the lid with a lever. If the spring is broken, a new one identical in height and thickness to the original should be obtained from a watch materials

Fig 3 Barrel and spring hooks.

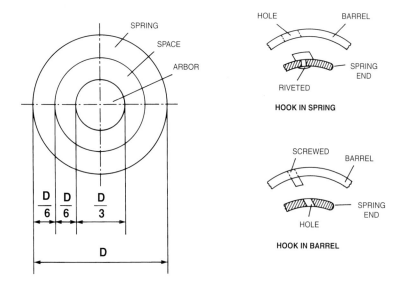

supplier. The length of the new spring should be such that when it is in the barrel it occupies half the space available. This is shown in Fig 3. If it is too long, it must be shortened and the new end softened by heating until red hot and then allowing to cool. A new hook and/or hole will be required at the outer end.

If a hook was fitted, then a new one must be filed from a piece of steel to suit the hole in the barrel. It must be made on the slant, as indicated in Fig 3, and then cut off and riveted to the spring through the new hole drilled in the end. If the spring is fitted to a hook attached to the barrel, the new hole in the spring must be cut at an angle in circular or rectangular form to mate with the barrel hook. This is also shown in Fig 3.

The state of the barrel hook should be examined in case it was the cause of the spring break. If it is worn, a new hook is made from a finely threaded steel wire and screwed into the barrel at the correct angle. If this is not possible with limited tools, then an old hole may be found on the barrel which can be broached out and a spring hook made to fit.

A new spring should be fitted using a spring winder. The amateur is unlikely to possess one, and unless a watchmaker will fit the spring, he will need to wind the new spring in by hand. This is adequate for working watches but could give erratic timekeeping. The hand-winding produces a spring which, when free, is slightly helical in form rather than a true spiral. This gives a component of force acting against the barrel lid, which tends to open it. If the lid is a poor fit, it may open and jam against the plate. This is unlikely, but the opening force will cause considerable friction between the spring and the barrel, resulting in poor timekeeping. When the new spring is tested, it may be found that the central hook on the arbor is not engaging with the hole in the spring. The spring centre should be bent gently inwards so that engagement is assured.

Repairing or cleaning a fusee chain

Fusee chains are often found to be broken, but provided both parts are sound they may be repaired. The chain should be soaked in benzine or releasing oil and then wiped clean. A close examination should then be carried out to make sure that the repair is worthwhile. Large amounts of rust would make repair pointless because breakage would occur again, but it may be possible to remove a faulty section.

A small punch is required, which may be made from a needle first softened or annealed by heating and then cooled slowly. The end should then be filed to the correct size and the needle hardened by heating to red heat and plunging into water; alternatively, an old staff may be adapted using the pivot as a punch. The broken ends of the chain are stuck in turn to a riveting stake with sticky tape so that the rivet to be removed is over a hole in the stake. The punch is used to produce male and female parts. These are then brought together over the hole in the stake. A new rivet is made from an annealed needle or steel pin by filing the end to the correct size. This is inserted firmly and the excess length removed from both sides with a thin file. The chain is replaced on the stake but not over a hole, and the ends of the new rivet are peened over with a punch to stop it falling out. The watchmaker's eyeglass will be required for most of these operations.

Fig 4 Fusee chain and hooks.

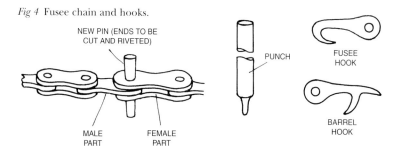

Should either hook from the chain be missing, a new one may be made from thin steel. The hole should be drilled (and if necessary broached) first and then the hook filed to shape. The shape can be seen in Fig 4, or another chain may be used as a pattern.

When the chain is deemed satisfactory by repair, or if it was not broken but merely required cleaning, it should be given a thorough soaking in oil; overnight is not too long. Then run the chain backwards and forwards over a groove in a smooth, rounded corner of wood, at the same time applying oil liberally so that all the chain joints are free and lubricated. Finally, wipe the chain clean with a benzine-soaked cloth.

Repairing a fusee click

If the fusee rotates in the winding direction satisfactorily, but fails to lock so that the spring tension immediately takes the chain back on to the barrel, it is likely that the fusee clicks are either worn or broken.

The fusee is shown in Fig 5. To strip the fusee, place the cone end in a boxwood stand and push out the pin, passing through the shaft and the blued-steel washer. Lift off the washer. The fusee may then be separated into the great wheel of brass, the maintaining-power ratchet wheel of steel with the two clicks or pawls, and the fusee cone with the driven ratchet wheel pinned to it. The clicks operate on this wheel, and examination will show if it is wear or breakage causing the slip. If it is merely rounding off of the click points, these may be filed or stoned to give satisfactory working, but excess wear or breakage will require a new click to be filed from steel. In extreme cases, the brass wheel with which the clicks engage may need some attention, although damage can usually be dressed out with a file.

Fig 5 Maintaining-power fusee.

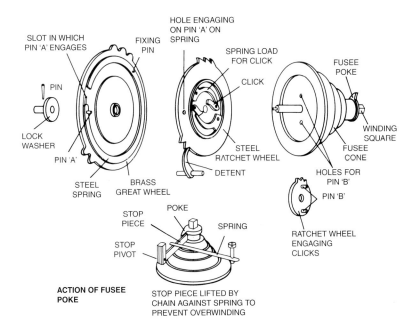

It is possible that the damage may not be confined to the clicks, but may also include the spring pushing the click into engagement. In this case, the fusee may be persuaded to operate with only one click, though it is better to file a new spring piece from steel. Both the clicks and the spring are secured by pins filed integral with the parts and peened over on the underside of the maintaining-power ratchet wheel. When the fusee is assembled with the new parts, the action should be tested before pushing the pin back through the blued-steel washer.

The older type of fusee without maintaining power has similar clicks in the construction which will cause slipping when worn. Repair is effected by similar methods.

In the event of a fusee being missing or damaged beyond repair, it may be possible to transfer one from another watch. The fusee must have the correct diameter and tooth spacing to mesh with the centre-wheel pinion. Differences in height can be accommodated by using a spacing bush to reduce end shake to an acceptable amount, provided the fusee stop and maintaining power detent will still engage.

Broken staff repairs

There are several ways to deal with a broken pivot on a staff: buy a new staff or make one, make a pivot repair with the re-pivoting tool (see page 136), or try the method below, which may be changed later by making a whole new staff. The author has used it successfully on several ocasions. It is also possible to repeat the operation if the new pivot is broken. In fact, it may be a good idea to try to avoid buying watches with a broken balance-staff pivot, but if the watch is important to you, you can always just accept that the watch is not working, or have it repaired by a skilled watch repairer, rather than attempt the repair yourself.

When pivots are broken on staffs that are difficult to replace or too expensive to be made economically, it is possible to replace half a staff by using the hub of the wheel on the staff as a joining sleeve. Fig 6 opposite shows the principle involved. This method is more useful if the broken pivot is not on the pinion half of the staff, but even on this half it is possible to wait until a suitable pinion with a good pivot is obtained before the repair is made. An alternative method is suggested on page 155 for pinion ends.

Staffs are removed from wheels and levers by using a punch and a stake. The punch must have a hole of suitable diameter to go over a pivot and deliver the blow to the shoulder of the staff. Many staffs have a taper and can only be removed in one direction; the dimensions should therefore be checked with a micrometer to be certain of the correct removal direction. The height of the wheel on the staff should also be measured to make sure that it is replaced correctly. If a replacement-part staff is found on which the pivot is of the correct diameter, the staff diameter may be reduced by careful filing if it is too large, or it may be set in with a non-permanent glue if it is too small. Always modify the replacement part rather than the wheel hub or the pivot hole in the plate. Finally, the total height of the composite staff must obviously be correct.

Fig 6 Staff repairs.

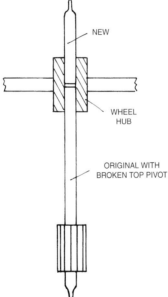

When satisfied with the repair, the 'new' staff should be tested for engagement in the depthing tool *set at the distance of the appropriate pivot holes in the plates.* When satisfactory, the repaired part should be checked for height between the plates. The watch can then be reassembled and given a trial run. (Do not wind fully because if the trial is not satisfactory, the movement will need to be dismantled again.)

Verge top pivot, bottom pivot, verge and crown wheel

The top pivot of the verge may be repaired in a similar fashion to the broken staff. It may be found that there is very little room for this technique, as the verge pivot is close to the balance. In this case, the brass boss on which the balance is fitted and the broken staff may be bored out to allow a new piece of staff to be fitted. This can be done with a hollow cutter, which removes the soft brass and allows the thin steel of the verge to be broken off. Fig 7 shows the sequence of events in this repair.

An alternative is to remove the balance, file away half the thickness of the balance from the boss and broken verge, and bond on a new piece of staff. This method is not as satisfactory as the first but is considerably easier to achieve. In either method, great care must be taken to avoid damage to the rest of the verge.

If the bottom pivot is broken, the correct method is to make a new verge. This can be done with a file, but it is a long process

Fig 7 Verge top-pivot repairs.

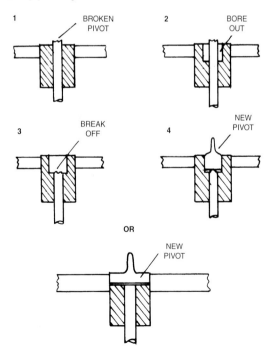

with a chance of breaking the new part at a late stage. Should the broken pivot be to hand, it is possible – with great patience and several tries – to bond the pivot back into place with epoxy resin glue or some similar immensely strong adhesive. This technique may also be used with the top pivot. The result may not be satisfactory for permanent running, but it is a method of producing a ticking or working-order verge watch. It may also be used with a replacement pivot, but this is more difficult because the broken staff and new pivot do not 'key' together.

If the verge is broken between the pallets, or the crown-wheel staff outer-end pivot is broken off, then a thin piece of bushing tube may be used to join the two halves or, in the latter case, to join the crown-wheel end to a new outer-end piece. There is very little room in this area for bulky extras, and epoxy resin glue may be needed to add strength. The inner pivot on the crown wheel presents a different problem, which may be solved by using epoxy resin glue to bond on a new pivot. An alternative is to use the method suggested below.

Generally speaking, a broken verge escapement is 'bad news', and specialist advice should be sought. Everything then depends on what the watch offers in terms of value, interest, and so on.

New pivot on the pinion end of a staff

If the pivot at the pinion end of a staff is broken and it is desirable or essential that the original pinion be retained, there may well be a problem. If there is sufficient length and room a new pivot can be bushed on to the old staff, as in the case of the verge above. However, there is often no room for a bush, for example, at the inner end of the crown-wheel staff in a verge watch. In this case, the only alternatives which retain the original pinion are bonding with epoxy resin glue or drilling out. Success with such glues have meant that the drilling method has not been personally tested. In principle, however, a new pivot could be let into the pinion end of a staff if, after softening and filing off the broken parts flush with the pinion face, a hole were drilled. Much will depend on the pinion-core diameter, but with a reasonable diameter available it is well worth trying if no other method can make the watch tick. The author has seen it done by a professional repairer, and in this context it should be borne in mind that there are watches for which it is worth paying a professional to carry out such difficult tasks. Here it is assumed that the collector has a watch of little value for which he is seeking a method of repair that will give him satisfaction. Depthing should be checked in the tool.

Straightening wheels, shafts and pivots

If a wheel is bent, it may be straightened by pressing between two flat surfaces. Some heat may be applied to the surfaces to ease the flexing. After this process, it is likely that there will still be a kink at any point where the original bend was particularly sharp, and this may be removed by light taps with a small punch. The teeth should not be hit in case they spread and interfere with meshing. If the wheel is broken as well as bent it may be soldered at the fracture. The repaired wheel should be put into its pivots and examined for truth, and then tested with its pinion and any tight spots eased. If any teeth are missing, a new piece of brass can be let into the rim of the wheel and soldered. Fresh teeth are filed to give satisfactory meshing when tested in the watch plates or in the depthing tool with the engaging pinion, *set at the distance apart of the relevant holes in the plate.*

Shafts may be straightened by tapping with the pane of a small hammer (or a similar-shaped piece of metal) on the concave side. If the high convex side is hit, damage is likely to occur. Heat applied to the anvil which will penetrate the shaft may help the process.

Pivots can sometimes be straightened. When the pivot has been softened, the straightening should be attempted in small increments. There is a good chance of the pivot breaking, but if the watch did not work in the first place because of the bent pivot there has been no loss.

In both shaft- and pivot-straightening operations, there is no point in pursuing perfection and incurring the risk of damage if the collector is satisfied with a nearly straight part which will allow the watch to tick. As soon as this is thought possible, the train should be assembled and tested.

New wheels

If a wheel is so badly damaged that repair is impossible, or if a wheel is missing completely, it is possible to transfer a suitable replacement from another movement. The number of teeth and the diameter is known or can be calculated by counting the teeth and measuring the size of the meshing pinion and the distance between the two shafts. The correct meshing of teeth is important, but if it is a case of making a watch go, then trial and error can be used.

If a wheel appears to be a satisfactory substitute, it should be mounted on the staff and a trial made with the appropriate pinion in the depthing tool (or the watch plates). If the wheel is too small, it may be possible to spread the teeth outwards by tapping with a hammer. The tooth sides will probably need easing with a file after this treatment. If the wheel is too large, the diameter may be reduced with a file. Again, the tooth sides may need easing. All modifications should be made to the replacement wheel, not the correct watch pinion, so that in the event of failure the watch is still intact and a fresh start can be made. Once the wheel runs satisfactorily with its pinion, the train should be assembled and tested before the whole watch is rebuilt.

If it is an escape wheel that is missing, then the problem is more difficult. With cylinder, duplex or detent escapement, the task is one for a professional antique watch repairer. The amateur might be able to make or adapt a crown wheel for a verge watch. A careful examination of a complete verge escapement will make the scope of the work clear, and the Bibliography for Chapter 5 contains books giving full details. This is a task for patient filing.

With a lever watch it might be possible to take the escape wheel from a scrap movement and fit this in the damaged watch, but if a wheel of the correct size cannot be found, then both the escape wheel and the lever may be planted in the damaged watch. The lever may need some adapting at the fork end. This transfer should only be attempted in a watch of little value, for new pivot holes will have to be drilled in the plates to accommodate the new escapement: this is an irreversible step. This technique was used by professional watchmakers in the latter part of the nineteenth century, when cylinder or duplex watches had their escapements changed to lever, but it has not been personally tested.

Filing out new parts

The file should never be underestimated. Provided that the work can be held in some way or a jig made to support it, and provided that adequate patience and time are available, a considerable range of replacement parts can be filed out. Bearing in mind that the conversion of a broken watch to a ticking watch is a considerable achievement, the fact that a part made by amateur hand-filing may leave much to be desired in timekeeping or continuous reliable running is trivial. As experience is gained, hand-made parts will improve in quality and it may be possible to return and improve earlier work. Parts the author has made in this way are a seven-tooth pinion, hands, clicks and hinges, and parts made by other collectors include balance wheels, cocks, staffs, a verge-bearing bracket and a verge. So when you are confronted with a seemingly impossible task, consider the possibilities of a file.

Bushing

If the pivot holes in a plate are so badly worn that bushing is required, then a decision has to be made. Bushing will require holes in the plate to be enlarged, which is an irreversible step. If the watch is of no great value, then bushing can be attempted without fear. The bushes can be purchased from a watch materials supplier, and come with an extended piece which can be broken off after the bush is fitted. The old hole will need to be broached out to give a good, tight fit on the bush. However, bushing should be regarded as a habit to be avoided unless it is absolutely necessary.

An alternative bad practice which is sometimes in evidence on watches is to spread the metal around the pivot hole by using a punch around the perimeter. The hole is then broached to size. This again is a habit to be avoided, but it cannot be denied that it may achieve some success.

Pivot holes which are jewelled may also need repair. Replacement jewels are available from watch materials suppliers, but their design is different from the jewels which were set in with screws. If the original jewel is complete but damaged, it is possible to effect a repair with epoxy resin glue which will enable the watch to run. It is also possible to transfer period jewels from an old movement.

Impulse jewels

These are sometimes found broken in an otherwise perfect lever watch. If a replacement jewel is not available, then a piece of steel wire may be used. The pin or jewel is not normally circular but ellipse or D-shaped, as shown in Fig 11 on page 63. The D-shape is the easier to form, but the hole in the roller should be the guide to the shape required. The D should be formed by removing one-third of the circular profile, leaving a flat surface. Some early lever watches appear to have been fitted with circular jewels and watches certainly function quite adequately with such jewels.

The replacement pin should be a sliding fit in the hole in the roller and the length just adequate to operate the lever without fouling the bottom pivot. The hole in the roller is filled with a non-permanent glue (such as shellac or varnish). The new pin is introduced and adjusted, and then the whole system put aside so that the glue sets.

Motion work

It is disappointing to find that all, or part of, the motion work is missing from an otherwise complete watch, but it is not an uncommon fault. If one wheel only is missing, it is possible to find a correct replacement in an old movement. The teeth numbers should be such that when combined with the two remaining wheels a 12 to 1 gear ratio is obtained. The diameter of the missing wheel must also be correct. The wheel centre may need broaching out to fit the pin or the cannon pinion on which it revolves.

In the absence of a suitable matching wheel, it is better to take a complete set of motion work from another watch movement. This may require one hole to be made in the underside of the dial plate to take the pin on which the idler wheel rotates. It may also be necessary to broach out (having first softened) the centre of the replacement cannon pinion. The parts removed should be retained, so that when a suitable matching wheel is found they can be replaced.

Hands

Replacement hands are available from watch materials suppliers, but they will be modern, pressed-out hands suitable only for post-

1800 watches. Older hands, such as beetle and poker designs, were hand-filed from steel and this is how replacements should be made. First, drill the holes through the hands and broach or file the centre to the correct square or round size. Then file the hands to shape. These designs are not simple one-plane affairs, but are contoured in two dimensions. A contemporary pair of hands should be used as a pattern. Finally, polish, degrease (with benzine) and then blue the hands.

Blueing

Steel screws, hands and springs and are usually blued in pocket watches. To achieve successful blueing, first polish and degrease the parts in benzine. They must not then be touched by hand.

Place the parts on a tray and heat the tray with a spirit lamp. The heating will cause the polished steel to change through a range of colours – pale straw, dark straw, brown-yellow, yellow-purple, purple, dark blue, pale blue – until the steel is finally colourless. The technique of blueing is to stop the process so that the steel remains a dark blue with perhaps a tinge of purple. One of the main problems is to obtain a uniform colour, which means a uniform temperature. If the tray is filled with brass filings and tapped during the process, this will facilitate the even distribution of heat. It is also possible to use a heated metal block for thin objects and to slide the part about to achieve the uniform colouring.

If the result is not satisfactory, the part can be made colourless by re-polishing and repeating the process. Acids can also be used to remove blueing, and if a modern acidic agent is used to remove or inhibit rust on a blued part it will emerge colourless, requiring reblueing. Acid treatments should be avoided.

A gun collector showed the author a special chemical fluid called 'Birchwood Casey "Perma Blue" liquid gun blue' which was used on blued-steel cases. However, such treatments should be employed with caution.

Epoxy resin glue

This is a two-component adhesive which is exceedingly strong when set. At room temperature the setting time is quite long, but it can be shortened considerably by placing the bonded parts on a radiator.

There is also a rapid-setting variety. Thus it is a very useful material and may be used to repair parts which would otherwise need to be replaced. Examples of its use are the bonding together of a shaft and pivot, the repair of a broken wheel, the repair of jewels and casework repairs.

Because of its great strength, it is not always possible to rebreak a part that has been joined with epoxy resin glue. This would not be true in the case of a pivot, where the area for bonding is very small, but if there is a large area then such a repair should be regarded as permanent, and it should only be used in situations where a permanent joint is required.

Case work

Damaged watch cases must be repaired with the appropriate material. It is against the law to use ordinary lead solder on hallmarked silver or gold. The silver used in watch cases is normally 925 parts of silver per thousand, the balance being copper. The remainder of this section is concerned with silver case work.

Silver soldering requires either a butane blow-lamp or mouth blowing from ordinary gas. Care must be taken to avoid melting adjacent joints, and the lowest-melting-point solder available should be chosen. Wet rags placed on adjacent joints will help to keep them cool. Parts to be soldered should be clean and grease-free and the correct flux should be used. Borax crushed in water is the usual material.

New pendants can be cast from old case metal, but the melting temperature of the silver is just too high for the small butane lamp, and a large torch or kiln is better. New bezels can be made from silver strip, but considerable skill is required to achieve a satisfactory result. New backs can be adapted from old cases. The best way to tackle the problems of major case work is to attend a jewellery class at a local school of art and craft. They will charge for the course and for the materials used, but the skills of a professional silversmith will be available.

For movements without cases it is possible to adapt cases without movements, or to make cases from brass, aluminium or perspex. This form of collecting is not of universal appeal, however, and if a case is to be made, silver is considered to be the most attractive material to use.

Glasses

Old watch glasses should be purchased if they become available. Unfortunately, the glasses will often be very small or very large. Watch materials suppliers or watchmakers may have some modern glasses, and these should be fitted to keep dust from the movement and to avoid damage to the dial or hands. If nothing else is available, it is sensible to provide protection with a plastic 'glass'.

The older type of glass fitted to watches before 1800 which had a flat in the centre (bull's-eye glass) is less common, and should be purchased and held as stock against future needs. Watches which should have these glasses are often found with a more modern glass fitted: these glasses will still have a very high dome and should not be discarded lightly, but they should be replaced with the correct glass and stored for later use.

Dials

Dial repairs are rarely completely satisfactory. All fillers seem to have a different colour or texture to the original dial. An alternative and interesting approach to a watch with a badly damaged dial or without a dial is to make a new one. Again the local school of art and craft will be able to help at the jewellery class where enamelling is practised. It is, however, possible to enamel successfully away from the class, because the materials are cheap and the temperatures involved are within the range of the small butane blow-lamp.

The new dial should be made from copper sheet – about 25 gauge for domed dials and 20 gauge for flat dials (which tend to flex and crack the enamel if made of thinner metal). Mark out the dial, cut to size and drill for hands. Annealing is achieved by heating to red heat and then plunging into cold water. Use a small bolt to hold the dial in the chuck of a hand drill held in a vice, and press the dial to a dome shape as the drill rotates. Use a piece of wood to bring pressure to bear on the dial, and annealing may be required part-way through the doming process. Smooth and clean the domed blank to a good finish, and then place in dilute sulphuric acid for a quarter of an hour. After this, only handle it with tweezers to avoid grease contamination. Wash the acid-treated blank with water. The acid process can be omitted provided that the blank is completely free of scale and grease.

Coat the blank with gum tragacanth (mixed from powder to paste with methylated spirits and diluted with water: 15g (½oz) gum to 1 litre (2 pints) water), place it on a coarse metal mesh and sieve the white enamel powder through a 60-gauge mesh on to the face. Treat both sides to avoid edges, but the butane method burns enamel off the back. Apply the butane lamp to the back of the dial until the enamel fuses at about 820°C (1500°F). Repeat this process until a satisfactory white finish is obtained.

Next, draw or paint the numbers with black painting enamel mixed with water to a suitable consistency. Considerable trial and error will be needed until satisfactory draughtsmanship is achieved. The black enamel can be washed off and a new start made as often as required. It is best to make a rotating table so that circular rings can be achieved with a fixed 'pen'. Similarly, some sort of radial jig is needed to insert the numerals. Roman numerals are easier to paint than Arabic (1,2,3...12).

When satisfied, the painting enamel must be fired. Painting enamel fuses at a lower temperature than the white enamel (about 730°C (1350°F)) so there is no problem with the white base. The dial is again heated from the back, for at no time should the lamp be played directly on to the face. Sometimes the enamel ground will crack at this late stage in the treatment, but in some cases it is possible to salvage the work by fusing new layers of white on top of the numbers and then painting on new ones.

New feet can be fitted to the back of the dial using epoxy resin glue to bond them on or, if preferred, they can be brazed on before enamelling. Enamel repairs to dials with chips are not satisfactory, for the enamel does not have the same whiteness as the original and the butane treatment tends to damage the remaining good enamel of the original. The original dial should be put to one side and preserved as part of the watch and a complete replacement made.

Another approach to dial repair which has not been personally tested but which may have possibilities is to use dentist's materials. Colour matching to teeth is certainly possible, so that matching to a dial should also be feasible. Consult your dentist.

Technical glossary

Arbor A shaft or axle.

Arcaded dial A dial on which the minute ring has arches between the numerals rather than a circular form.

Automatic Self-winding.

Auxiliary compensation Additional compensation added to a bimetallic balance to reduce middle-temperature error.

Balance The oscillating spoked wheel which controls the rate at which the mainspring is allowed to unwind.

Balance spring The spiral or helical spring controlling the balance vibration.

Balance staff The axle on which the balance is fitted.

Banking A system to control the arc of vibration of a balance or, in the lever watch, the motion of the lever.

Barrel The cylindrical container for the mainspring.

Beat The audible tick of the watch. A watch that is in beat has an even tick.

Beetle hand The hour hand used on eighteenth-century watches in combination with the poker minute hand.

Bezel The ring-shaped piece of case holding the glass.

Bimetallic balance A balance wheel whose rim is made of two metals such that the differential expansion rate counteracts the effects of temperature changes on the rate of the watch.

Bow The hanging ring of a pocket watch.

Bull's-eye glass A high-domed watch glass with a flat centre piece.

Button The winding knob of a keyless watch.

Calibre A term used to denote the size (and shape) of a watch.

Cam A contoured shape which rotates to give a special motion to a follower resting on the cam.

Cannon pinion The pinion driving the motion work with a long hollow arbor, which fits over the extended centre-wheel shaft between the dial and the plate.

Centre wheel The second wheel of the train, rotating once per hour.

Chaffcutter A Debaufre-type escapement, described on page 56.

Chapter ring The ring marked on the dial with hour divisions.

Chinese duplex A duplex escapement watch with double-locking teeth so that two complete balance vibrations are required for escape. The watch advances in increments of one second.

Chronograph A watch with a seconds hand that can be started, stopped and reset independently of the mean-time hands.

Click A pawl or detent inhibiting motion in one direction.

Club-foot verge A Debaufre-type escapement.

Club-tooth lever A lever escapement with the type of escape wheel described under the Swiss lever on page 64.

Cock A bracket supporting the pivot of a wheel. Usually the term refers to the balance cock for the top balance pivot.

Compensated balance A bimetallic balance designed to counteract the effect of temperature change on the rate of a watch.

Contrate wheel A wheel with teeth at right-angles to the plane of the wheel. The fourth wheel in a verge watch is an example.

Coqueret A hard steel bearing on the balance cock of continental watches.

Crank lever escapement A Massey lever escapement described on page 60.

Crank roller escapement Another name for the crank lever escapement.

Crown wheel escapement The verge escapement described on page 51.

Curb pins The pins attached to the regulator which loosely hold the balance spring so that its working length can be varied as the regulator is adjusted.

Cylinder escapement The first successful alternative to the verge escapement introduced in 1726. It is described on page 52.

Dart The safety pointer below the lever fork in the double-roller lever watch.

Dead-beat escapement An escapement without recoil.

Dead-beat verge A Debaufre-type escapement.

Debaufre-type escapement An escapement based on an invention by Peter Debaufre in 1704; it is described on page 56.

Deck watch An accurate watch used aboard ship during astronomical observations; it is usually contained in a wooden box.

Depth A term to describe the amount of penetration between two meshing gears.

Detached escapement An escapement in which the balance vibration is free from friction except during the unlocking and impulse action.

Detent A holding piece which stops movement in one or two directions.

Detent escapement The (pocket) chronometer escapement described on page 55.

Divided lift An escapement in which the lift-giving impulse is partly a result of pallet slope and partly of the escape-wheel tooth shape.

Double bottom case A case in which the back opens to reveal a second bottom pierced by a winding hole.

Double roller A lever escapement with two rollers. One roller is for impulse action and the second (smaller) one for safety action.

Draw The shaping of the escape wheel teeth so that the lever pallets are drawn into the escape wheel and on to the banking pins to prevent friction due to accidental lever motion.

Drop The free travel of the escape wheel between escape and locking.

Duplex escapement An escapement based on a design by Dutertre, described on page 53.

Dust cap A cover placed on the movement in key-wind watches.

Dutch forgery A term used to describe watches with a bridge-type balance cock, an arcaded dial with or without a scene and often a repoussé case. The work is of mediocre quality and marked with an English 'maker'. Probably made partly in England and partly on the Continent.

Ebauche An unfinished movement supplied by a factory to the watchmaker, who finishes and signs it.

End shake The axial clearance between a shaft and its bearings.

Endstone A disc-shaped jewel on which the end of the balance top pivot rests.

Engine turning The common form of nineteenth-century case decoration.

English lever escapement The escapement used almost exclusively by English watchmakers from 1850 until 1920, described on page 00.

Entry Pallet The pallet on a lever which receives impulse as the escape-wheel teeth enter.

Equation of time The relationship between solar time (based on the position of the sun) and mean time (based on averaged solar motion).

Escapement The part of a watch movement which constrains the train motion to small increments. It consists of the escape wheel, lever and balance.

Escape wheel The wheel in the movement connecting with the lever or balance.

Exit pallet The pallet on a lever which receives impulse as the escape-wheel teeth leave.

Figure plate The small dial indicating the amount of regulation, fitted to watches with Tompion regulation.

First wheel The great wheel on the fusee or going barrel.

Fourth wheel The fourth wheel of the train which rotates once per minute if a seconds hand is fitted.

Free sprung A watch with no regulator and curb pins. Regulation is achieved by the timing screws on the balance.

Frictional rest escapement An escapement in which the balance motion is affected by friction during the major part of the vibration.

Full plate A watch in which the top plate is complete or has a barrel plate. The balance is fitted on top of the plate.

Fusee The conical-shaped piece with a spiral groove for the fusee chain which equalises the mainspring torque.

Going barrel A spring barrel driving the watch train without the use of an intermediate fusee and chain.

Great wheel The first wheel of the train on the fusee or going barrel.

Greenwich time The local mean time at Greenwich, used as a basis for longitude and world time-zones.

Guard pin The vertical pin (behind the fork on the lever of an English lever escapement) which gives safety action on the roller.

Hairspring The balance spring.

Half plate A watch in which balance, lever, escape wheel and fourth wheel have separate cocks.

Hallmark The assay mark on English silver indicating date and quality.

Heart piece The cam used in a chronograph to reset the centre seconds hand.

Hog-bristle regulator Flexible bristles arranged to limit the arc of vibration of a pre-balance spring watch.

Horizontal escapement The cylinder escapement described on page 52.

Horns The forked end of a lever.

Hunter A watch case with a hinged solid cover over the glass. If fitted with a small glass it is called a half hunter.

Impulse The push given to the balance by the escapement.

Index The regulator pointer.

Isochronism The property of taking the same time for a balance vibration independent of the arc of vibration.

Jewels Bearings made of precious stones such as ruby. Modern jewels are synthetic.

Karrusel A rotating escapement designed to avoid positional error, described on page 71.

Kew Certificate A rating certificate (A, B and C grade) given by Kew Observatory from 1884.

Keyless A watch that is both wound and handset without a key.

Lancashire size An English scale for the size of a movement.

Lever escapement A detached escapement described on page 00

Lift The angular motion of the lever.

Ligne A continental unit of measurement of watch size.

Mainspring The spiral spring in the barrel providing power.

Maintaining power An arrangement in the fusee to keep power on the train while the watch is being wound. This avoids the watch stopping or faltering during the operation.

Massey lever escapement A detached lever escapement described on page 60.

Mean time The conventional time shown by clocks and watches based on average solar motion.

Middle-temperature error A residual timekeeping error in watches with a compensated balance.

Motion work The gearing under the dial used to make the hour hand rotate at one-twelfth of the speed of the minute hand.

Movement The watch works without case, dial and hands.

Oil sink The small depression in the watch plate around a pivot hole designed to hold oil in place.

Ormskirk escapement A Debaufre-type escapement described on page 56.

Overcoil The last coil of a balance spring, which departs from the spiral by being bent above the spring to give a better approach to isochronism. Invented by Breguet.

Pair case A watch with an inner case and a separate outer case. It was in general use until about 1800; uncommon after 1830.

Pallet The part of the escapement through which impulse is transferred from the escape wheel to the balance wheel or lever.

Passing crescent The indentation in the roller in a lever watch to allow the guard pin or dart to pass at the instant of impulse.

Pendant The part of the watch case to which the hanging bow is attached.

Pillar plate The plate of the watch nearest to the dial.

Pillars The distance pieces separating the watch plates.

Pinion A small steel gear wheel (6–12 teeth) driven by a larger brass wheel.

Pin-lever escapement An inexpensive lever escapement described on page 65.

Pivot The small-diameter part at the end of a shaft which is supported in a bearing.

Plates The flat brass discs supporting the train of the watch. The plates are separated by pillars.

Poker hand The minute hand which is used on eighteenth-century watches in combination with the beetle hour hand.

Positional error An error due to the variation in the rate of a watch in different positions: pendant up or down, etc.

Potence A hanging bearing, such as the lower balance pivot on a full-plate watch.

Rack lever escapement An early lever escapement described on page 59.

Ratchet wheel A wheel with saw-shaped teeth which will rotate in one direction. Rotation in the other direction is impeded by a pawl.

Rate The daily rate of loss or gain of a watch.

Recoil The backward motion of a watch when the escapement is unlocked.

Regulation The term used to describe the adjustment of the timekeeping of a watch.

Repeater A watch which gives audible indication of the approximate time by sounding gongs when a push piece is operated.

Rocking bar A device used to change from winding to handsetting mode (or vice versa) in keyless watches.

Roller The disc fitted on the balance staff in a lever watch to receive impulse and give safety action.

Roskopf escapement An early pin-lever escapement.

Safety roller The smaller roller for safety action in a double-roller lever watch.

Savage two-pin escapement An early lever escapement described on page 61.

Second wheel The centre wheel of the train.

Self-winding A watch with an eccentric weight pivoted so that it will always swing to the low position. The swinging motion winds the watch.

Set hands square The square on the end of the cannon pinion used to set the hands on a key-wind watch.

Set up The initial adjustment of tension in a spring.

Shifting sleeve A device used to change from winding to hand-setting mode (or vice versa) in keyless watches.

Single roller A lever watch with a single roller fulfilling both impulse and safety requirements.

Solar time The time indicated by solar position. A day is the time elapsed between two transits of the sun. This is not a constant.

Spade hand An hour hand with an enlarged end of similar shape to the spade symbol on a playing card.

Split seconds A chronograph with two independent seconds hands each of which can be operated separately.

Stackfreed An early regulation device based on the friction between a cam and a spring-loaded follower.

Stem wind Keyless winding and handsetting through a button.

Swiss lever escapement The Continental form of lever escapement described on page 64.

Table-roller lever escapement The English lever escapement described on page 63.

Temperature compensation Compensation for the changes in timekeeping caused by changes in temperature, usually by bimetallic balance.

Terminal curve The special end shape given to a balance spring to give isochronous motion.

Third wheel The third wheel of the train.

Threequarter plate A watch in which the balance, lever and escape wheel have separate cocks.

Timing screws The two (or four) screws at the ends of the balance arms (if fourscrews, also at right-angles to the ends of the arms). They are not compensation screws.

Top plate The plate of the watch furthest from the dial.

Tourbillon A rotating escapement designed to avoid positional error, described on page 71.

Train The series of meshing wheels and pinions connecting the fusee or going barrel to the escapement.

Up-and-down dial An extra indicator on a watch dial to show the state of mainspring winding.

Verge The vertical staff below the balance wheel carrying the pallets of the verge escapement.

Verge escapement Early escapement described on page 51.

Watch paper A circular piece of paper or cloth often carrying printed information or advertising, placed between pair cases by the seller or repairer. The paper takes up play and inhibits rubbing.

Wheel A large, brass train wheel which drives a smaller steel pinion.

Winding square The square end on the fusee or barrel arbor used for winding a key-wind watch.

Worm An endless screw which is rotated to turn a gear placed tangentially to the worm surface. The plane of the gear is that of the worm axis.

USEFUL INFORMATION

HALLMARKS

Hallmarks on silver watch cases can be used as a dating guide. After about 1700 (Plate Duty Act, 1719), silver cases of English origin, and those of European origin that have passed through an English assay office, will carry a hallmark indicating the metal and the place and year of assay. This system of marks may be used as a guide (reliable or otherwise) to the date of a watch. Watches can be recased and marks can be faked, so the information must be treated with care and used in combination with that obtained from other sources such as Baillie's *Watchmakers and Clockmakers of the World* (see Bibliography).

Casemakers' marks, usually consisting of initials, also appear on some cases, which may help in dating. The best guide to

Fig 1 Hallmarks.

LION PASSANT
(INDICATING STERLING
SILVER)

LEOPARD'S HEAD
USED IN ADDITION
TO CHESHIRE CITY
ARMS UNTIL 1838
(WITH CROWN
BEFORE 1823)

MAKER'S
INITIALS

DATE LETTER
INDICATING 1828

CHESHIRE ARMS
(INDICATING ASSAY
CENTRE)

**MARKS OF OTHER
ASSAY CENTRES**

LONDON UNTIL 1820

LONDON AFTER 1820

BIRMINGHAM

casemakers' marks is Priestley's 'Watch Casemakers of England 1720–1920' (see Bibliography).

European marks are not so easily interpreted, but may be used as a guide to the metal quality. The number 800 or 925 indicates the parts per thousand of silver in the case alloy.

On English hallmarked silver, the indication of sterling silver (92.5 per cent pure) is a lion passant. There were a number of assay offices but in the period of interest of this book the majority of cases were assayed in London, Birmingham or Chester. These centres are represented by a leopard's head, an anchor and the Chester arms respectively (see Fig 1).

Alphabetical letters used in sequence represent the date; however, they were not changed on 1 January but some time in mid-year, so that dating is not precise. The date-letter style in the sequence A to Z (with some letters occasionally missed out) lasts from 20 to 26 years and then the style is changed, so that with the assistance of an inexpensive book of silver marks or a good memory, dating should present no difficulty.

Marks are often worn and not very clear, but they should appear on each part of the case to obviate this difficulty (on pair cases, any difference in marks indicates a new or married case). Fig 1 shows the marks inside the case of the watch shown on page 126. The method of interpretation of marks is given in the annotations.

WATCHMAKERS

The list opposite is of surnames of some important post-1750 watchmakers. It is not exhaustive. These are names of which a collector should be aware, and if he finds a watch with such a name it would be worth investigating. Much more detail about the maker must then be found, and the starting place for this is that invaluable reference, Baillie's *Watchmakers and Clockmakers of the World* (see Bibliography).

It must be appreciated that there are often several makers with the same surname, and it is essential that the collector is certain about the watch if he is paying for the name. However, if the watch is offered at the market price for an anonymous or non-significant maker's piece, then purchase can be made in the hope that research will show a good buy.

The list is restricted to the post-1750 period since it is considered that *any* watch earlier than this is worth investigation. No details beyond surname are given so that the collector is not confused. The recognition of the name is the first important step.

Arnold	Gout	Pendleton
Barraud	Grant	Pennington
Beatson	Harwood (wrist)	Perrelet
Berthoud	Jurgensen	Philippe
Bonniksen	Kendall	Pouzait
(Karrusel)	Kullberg	Recordon
Breguet	Lecoultre	Reid
Brockbank	Lepine	Rentzsch
Cole	Leroux	Robin
Dent	Le Roy	Roskell
Dutton	Leschot	Savage
Earnshaw	Litherland	Tavan
Ellicott	McCabe	Ulrich
Emery	Margetts	Vulliamy
Frodsham	Massey	
Girard	Mudge	

WATCH SIZES

Watch movements of English makers from the second quarter of the nineteenth century onwards carry a size indication stamped on the underside of the dial plate. This consists of a single number between 0 and 40, possibly followed by two numbers written over each other, eg $12\frac{8}{8}$ Lancashire watch size, which gives the overall movement diameter. To interpret this number, size 0 represents 1in and the diameter is 1in plus the number of 30ths of an inch given by the number plus $\frac{5}{30}$ in of 'fall' (the allowance for the larger size of dial plate to allow the top plate to hinge into the case). Thus, in the example above the dial-plate diameter is $(1 + \frac{12}{30} + \frac{5}{30}) = 1.567$in.

The numbers written over each other give the pillar height. To interpret this, size $\frac{8}{8}$ is 0.125in, and if the top number is altered then the height is 0.125in *plus* the number of 144ths of an inch indicated by the top number. If the bottom number is altered, the height is 0.125in *less* the number of 144ths of an inch indicated by the bottom number. Thus, in the example above the pillar height is $(0.125 - \frac{2}{144}) = 0.111$in. The range of pillar heights is from $\frac{0}{6}$ to $\frac{30}{0}$ that is from 0.0833in to 0.333in.

Consideration of the history of these sizes would indicate that the sizes were for use with gauges rather than in the awkward decimals produced above and that the gauges were based on fractional figures obtained on a system based on 12ths. (See also *Antiquarian Horology*, 2, vol 8, March 1973, 203–4.)

Continental watch sizes are based on *lignes*, a system of units in which 1 ligne is 2.255 millimetres.

WEIGHTS AND MEASURES

In metric measurements, 1lb (pound) is 453.6g and 1in is 25.4mm.

Jewellers and silversmiths measure in troy weight in which:

24 grains make 1 pennyweight (dwt)
20 pennyweights make 1oz (ounce) troy
12oz troy make 1lb troy

Thus, 1oz troy is 480 grains and 1lb troy is 5,760 grains. In the normal avoirdupois system, 1lb is 16oz and 1lb avoirdupois is 7,000 grains, so that 1oz avoirdupois is 437½ grains.

WHERE TO SEE WATCHES

Large collections are usually kept in museums. Many towns and cities will have such collections, and those listed below have been visited, but it is not suggested that these are the only collections available; smaller ones are often contained in local museums and galleries. A letter to the local tourist centre or authority before visiting an unfamiliar district will help you to find out if such a collection exists. The Antiquarian Horological Society or the British Horological Institute might be useful sources of information for collections both in the UK and elsewhere. Membership of either or both of these organisations is well less worthwhile for the monthly (BHI) or quarterly (AHS) journals.

The list below is by no means complete, but it does record most of the larger or more interesting collections in the UK. The list also includes some collections in Europe and America. *It is advisable to check this data before making a special journey.*

Collections in London

British Museum, Great Russell Street, WC.

Guildhall Clock Room, Aldermanbury, EC.

National Maritime Museum, Greenwich, SE. This collection includes chronometers.

Science Museum, Exhibition Road, SW. Building work starting in 1999 means that the collection will be stored.

Victoria and Albert Museum, Exhibition Road, SW.

Libraries in London

British Library, St Pancras, NW.

Guildhall Library, Aldermanbury, EC.

Science Museum Library in City and Guilds College, London University, Exhibition Road, SW. This library is particularly useful for UK and foreign journals.

Science Reference Library and Patent Office Library, Chancery Lane, WC. Both will move to St Pancras in mid-1999.

There will be other libraries in the UK containing horological material: a local library should have a listing which includes their special interests. There is an inter-library loan service for books, for which there may be a charge.

Other collections visited

Basingstoke Willis Museum, New Street.

Bournemouth Russell-Cotes Art Gallery, East Cliff.

Bury St Edmunds John Gersham Parkington Memorial Collection, Angel Corner.

Cambridge Fitzwilliam Museum, Trumpington Street.

Canterbury Beaney Institute, High Street.

Coventry Contact the Local History Library for information.

Exeter Royal Albert Memorial Museum, Queen Street.

Hove Museum of Art, New Church Road.

Lincoln Usher Gallery, Lindum Road.

Liverpool Merseyside County Museum, William Brown Street. See also Prescot.

Oxford Museum of the History of Science, Broad Street.

Prescot Prescot Museum of Clock & Watchmaking, 34 Church Street, Snowshill Manor, *Broadway*.

Upton (near Newark) BHI collection.

Other locations housing at least 20 watches
Contact the local library or tourist centre for details.

Alton	Dover	Northampton
Aylesbury	Dumfries	Norwich
Bath	Dundee	Preston
Belfast	Edinburgh	St Albans
Birmingham	Hastings	Salisbury
Bradford	Hereford	Sunderland
Bristol	Ipswich	Taunton
Burnley	Keighley	Tunbridge Wells
Carlisle	Leeds	Waddesdon
Chelmsford	Leicester	Warrington
Cheltenham	Newbury	York
Chester	Newcastle	

European collections visited

France

Paris Conservatoire Nationale des Arts et Metiers.
    Petit Palais

Switzerland

Basel La Chaux de Fonds, Musée Internationale d'Horlogerie.
    Haus zum Kirschgarten, Elisabethenstrasse 27–9.

Germany

Furtwangen Deutches Uhrenmuseum. Mainly Black Forest clocks.
Schramberg Staatmuseum.
Schwenningen Heimatmuseum.
Stuttgart Wurttembergisches Landesmuseum. Mainly sixteenth-
    and seventeenth-century clocks.
Wuppertal Abeler Museum.

US collections (not visited)

Bristol, Connecticut American Clock and Watch Museum.
Columbia, Pennsylvania National Association of Watch and
    Clock Collectors, Inc.
New York City Metropolitan Museum of Art.
Rockford, Illinois Time Museum.
Washington, DC Smithsonian Institute.

Patents

The Patent Office Library in London (see earlier note regarding the move to St Pancras) keeps both British and foreign patents, which may be viewed. Some photocoyping is possible, but it is also possible to obtain copies direct from Patent Office Sales, Unit 6, Nine Mile Point, Cwmfelinfach, Cross Keys, Gwent, NP1 7HZ (tel 01633 814000).

Other sources

UK

Aked, C. K., *Complete List of English Horological Patents up to 1853* (Brant Wright Associates Ltd, 1975).

Patents for Invention, Abridgements of Specifications, Class 139, Watches, Clocks and Timekeepers, 1855–1930. Limited edition facsimile reprint, a copy of which is kept in the Patent Office Library – see above.

US

Eckhardt, G. H., *United States Clock & Watch Patents, 1790–1890,* (US, 1960.)

Townsend, G. E., *Encyclopedia of Dollar Watches,* (US, 1974).

UK Patent Office.

Switzerland

Bibliothèque Nationale, Berne, Switzerland.

UK Patent Office.

WHERE TO BUY WATCHES

Markets

London

Bermondsey, Friday morning (early), London Bridge station.

Islington, Saturday, Angel station.

Portobello Road, Saturday, Notting Hill Gate station.

Many 'general' markets in other towns have watches on some stalls.

Watch Fairs

Entry fee usually applies. Two examples are given below, but details of fairs are contained in *Antiquarian Horology, Horological Journal* and *Clocks* (a monthly magazine from newsagents).

Brunel Clock & Watch Fair, Brunel University, Kingston Lane, Uxbridge (near London).

Midland Clock and Watch Fair, National Motorcycle Museum, Solihull.

Auction sales

The main auction houses – Bonham's, Christie's, Phillips and Sotheby's – have specialist auctions of watches and clocks, at which prices will be representative of the perceived value of the watch. These companies also have associated 'branches' in provincial auction rooms. Other auctions and house clearance auctions also offer watches for sale, but the purchaser will need to view the watches to assess them and either attend the auction or leave a bid with the auctioneer.

Postal auction

A postal auction enables a bid to be made based on the catalogue. Contact D. Penney, Groom's Cottage, Elsenham Hall, Elsenham CM22 6DP (tel 01279 814946).

Postal fixed-price catalogue sales

The catalogue gives a full description and illustration(s) of the watches on offer. Contact Pieces of Time catalogues (tel 0171 6292422).

Both the postal auction and postal fixed-price sales include a provision for customer satisfaction, but the author has never used these methods.

Antique shops, watch retailers, jumble sales and car boot sales will all offer watches in various states.

WHERE TO BUY WATCH MATERIALS

The author has no experience with any of the firms listed below. All the information has been extracted from advertisements in publications such as *Horological Journal, Antiquarian Horology* and *Clocks* magazine. These randomly selected advertisers, and others, offer services for repairs and spare parts.

Watch materials H. S. Walsh & Sons Ltd (tel 0181 778 7061).

C. R. Frost & Son Ltd (tel 0171 253 0315).

Watch glasses Euro Glasses (tel 01205 722474).

Watch dial repairs Lynton Dials (tel 01328 863666).

Watch repairs Roy Wadsworth (tel 01283 704018).

Jon Van de Geer (tel 01722 412841).

R. A. James (tel 01283 814596).

There should also be local watch repairers in most areas and a 'personal repairer' would be a valuable asset. The reader must make his own decisions in this matter.

Courses in watchmaking and repair

For information on courses, contact the British Horological Institute. The BHI itself offers courses, but would also be able to supply information about professional courses in other approved locations. For example, Hackney College in London normally offers watchmaking courses.

For courses in enamelling and silver work, contact your local college of art.

SOCIETIES AND INSTITUTIONS

Antiquarian Horological Society. Quarterly Journal. For details, contact The Secretary, The Antiquarian Horological Society, New House, High Street, Ticehurst, Wadhurst, East Sussex TN5 7AL tel. 01580 200155.

British Horological Institute. Monthly journal. For details, contact British Horological Institute, Upton Hall, Upton, Newark, Nottinghamshire NG23 5TE tel 01636 813795/6.

Both these groups produce interesting journals and hold local branch meetings and lectures. The AHS is mainly concerned with history and the BHI with practical work, current technology and some history.

National Association of Clock and Watch Collectors Inc, PO Box 33, Columbia, Pennsylvania 17512, US.

BIBLIOGRAPHY

General

Good, R., *Britten's Watch and Clockmaker's Handbook, Dictionary and Guide* (16th edition, 1978).

Chapter 1

Camerer Cuss, T. P., *The Country Life Book of Watches* (1967) (or Cuss, Terence, *The Camerer Cuss Book of Antique Watches* (1976)).

Gould, R. T., *The Marine Chronometer* (1960).

Jaquet, E. and Chapuis, A., *Technique and History of the Swiss Watch* (1970).

Rees, A., *Clocks, Watches and Chronometers* (1819, reprint 1970).

A comprehensive survey of keyless winding history researched by V. Mercer may be found in the *Horological Journal* in six parts (Volume 127: No 3, September 1984; No 5, November 1984; No 8, February 1985. Volume 128: No 5, November 1985; No 12, June 1986. Volume 129: No 7, January 1987).

Chapter 2

Billeter, C., *Le Réglage de Précision* (Bienne, 1930).

Chamberlain, P. M., *It's About Time* (1964).

Clutton, C. and Daniels, G., *Watches* (1965).

Cutmore, M., *The Pocket Watch Handbook* (1985).

Gazeley, W. J., *Clock and Watch Escapements* (1975 reprint).

Haswell, J. E., *Horology* (1975 reprint).

Rawlings, A. L., *The Science of Clocks and Watches* (1980 reprint).

Weaver, J. D., *Electrical and Electronic Clocks and Watches* (1982).

Chapter 3

Bacon, D.H., 'Watch Production in English Factories', 1870–1930; *Antiquarian Horology*, 23, 2, Winter 1996, pp. 117-133.

Carrington, R.F. and R.W., Pierre Frédéric Ingold and the British Watch and Clockmaking Company, *Antiquarian Horology*, 10, 6, Spring 1978, pp. 698–714.

Cutmore, M., *Watches 1850–1980 (1989)*.

Harrold, M. C., 'American Watchmaking 1850–1930', *NAWCC Bulletin Supplement*, 14, Spring 1984.

Horological Journal, 77, April 1935, pp.254–7; High Grade Watches made in London.

Jaquet, E. and Chapuis, A., *The Technique and History of the Swiss Watch* (1970).

Landes, D. S., *Revolution in Time* (1983).

Marsh, E. A., *Watches by Automatic Machinery At Waltham* (1896, reprint 1968).

Weiss, L., *Watchmaking in England 1760–1820 (1982)*.

Books covering similar ground to the Flume *Werksucher* include: Jacob, G., *Ein Blickgenügt Werk-Erkennung Gesamtausgabe* (1949); *Catalogue Officiel des Pièces de Rhabillage pour Montres Suisses* (1949); and *Bestfit III Encyclopedia of Watch Materials, Part 1* (US,1961).

Chapter 4

Baillie, G. H., *Watchmakers and Clockmakers of the World* (1963 reprint).

Cutmore, M., *Pin-Lever Watches* (1991).

Loomes, B., *Watchmakers and Clockmakers of the World*, Volume 2 (1976).

Chapter 5

Britten, F. W., *Horological Hints and Helps* (1977 reprint).

De Carle, D., *Practical Watch Repairing* (1946 edition).

Fried, H. B. *Repairing Quartz Watches* (1983).
          Watch Repairer's Manual (1986).

Glasgow, D., *Watch and Clockmaking* (1885, reprint 1977).

Saunier, C., *Watchmaker's Handbook* (1892, reprint 1990).

Swinkels, B., *Enamelling* (Hale 1975).

Appendices: Hallmarks

Priestley, P. T., 'Watch Casemakers of England 1720–1920', *NAWCC Bulletin Supplement*, 20, Spring 1994.

ACKNOWLEDGEMENTS

The production of this book has involved consultations with friends, libraries, museums and watch dealers. All have been helpful. The photographs are from various sources including Seiko and Newmark, W. R. Milligan and my own, less professional products. The line drawings are from various sources including Edan Art, David & Charles and myself. The acceptance by my wife that she is married to a madman has been long appreciated.

INDEX